The TAB Book of Arduino Projects

The TAB Book of Arduino Projects

36 Things to Make with Shields and Protoshields

Simon Monk

New York Chicago San Francisco
Athens London Madrid
Mexico City Milan New Delhi
Singapore Sydney Toronto

McGraw-Hill Education books are available at special quantity discounts to use as premiums and sales promotions or for use in corporate training programs. To contact a representative, please visit the Contact Us page at www.mhprofessional.com.

The TAB Book of Arduino Projects:
36 Things to Make with Shields and Protoshields

1 2 3 4 5 6 7 8 9 0 DOC/DOC 1 2 0 9 8 7 6 5 4

ISBN 978-0-07-179067-3
MHID 0-07-179067-5

This book is printed on acid-free paper.

Sponsoring Editor
Roger Stewart

Editing Supervisor
Donna Martone

Production Supervisor
Pamela A. Pelton

Acquisitions Coordinator
Amy Stonebraker

Project Manager
Patricia Wallenburg, TypeWriting

Copy Editor
James K. Madru

Proofreader
Claire Splan

Indexer
Claire Splan

Art Director, Cover
Jeff Weeks

Composition
TypeWriting

To my son Matthew

About the Author

Dr. Simon Monk (Preston, UK) has a degree in cybernetics and computer science and a Ph.D. in software engineering. He spent several years as an academic before he returned to industry, co-founding the mobile software company Momote Ltd. He has been an active electronics hobbyist since his early teens and is a full-time writer on hobby electronics and open-source hardware. Dr. Monk is the author of numerous electronics books, specializing in open-source hardware platforms, especially Arduino and Raspberry Pi. He is also co-author with Paul Scherz of *Practical Electronics for Inventors*, Third Edition.

You can follow him on Twitter, where he is @simonmonk2.

Contents

Acknowledgments

I am very grateful to my son Stephen, the musician of the family, for his help with the "Sound and Music" section of the book.

Introduction

This book contains 36 Arduino projects. Some are easy to make, whereas others require some expertise with a soldering iron. You do not need a strong grounding in electronics engineering to build these projects. Although you will find some theoretical explanations, this is primarily a book that shows you in detail how to make the Arduino projects that it contains.

Some experience with a soldering iron will be helpful. The only tools you will need are screwdrivers, pliers, snips, and a soldering set.

Arduino

Arduino has become the most popular open-source hardware for building your own microcontroller projects. There are many reasons for this:

- Low cost ($25 or less)
- Cross-platform (you can use it with PC, Mac, or Linux)
- Simple to program
- Ready-made plug-in shields that add hardware such as Ethernet, liquid-crystal display (LCD) screens, and so on
- Active and helpful community

Arduino Boards

Probably the two most common Arduino boards in use are the Arduino Uno and the Arduino Leonardo (Figures I-1 and I-2).

There are pros and cons for each board. The Leonardo is a little cheaper and can do some tricks, such as impersonating a USB keyboard or mouse, that the Uno is not capable of. However, the Leonardo is a newer device, and there are some

FIGURE I-1 Arduino Uno.

FIGURE I-2 Arduino Leonardo.

computability problems with older shields and libraries. However, most of the projects in this book will work with the Leonardo. In fact, a few of the projects in this book will only work with the Leonardo because they use the USB keyboard impersonation feature. Refer to the start of each project to check for compatibility or otherwise with different Arduino boards.

The Arduino Uno is a more common board. It is more expensive, but it does have a removable processor chip, which gives the advantage that should you accidentally short an output pin and destroy the processor chip, you can buy a new one for a few dollars. If you do that to a Leonardo, you will need to buy a new Leonardo.

If you get advanced in your Arduino project making, you can take a programmed processor from an Arduino Uno and build it onto a custom printed circuit board (PCB) or stripboard and then replace the processor chip with a new one for the next project. Again, this is not possible with the Leonardo.

If you are buying an Arduino for this book and you have no older Arduino kit, then I would start with a Leonardo. You may find that you end up buying an Uno if you catch the bug!

This book uses the Arduino R3 and the Leonardo. Although older versions of the Arduino should work, versions prior to R3 have fewer sockets around the edge, so an Uno R3 or Leonardo is recommended.

As you can see from Figures I-1 and I-2, both Arduino boards have a similar layout with connector strips down each side and a USB socket at one end. The Uno has a big B-type USB connector, whereas the Leonardo has a micro-USB connector.

All the projects in this book require an Arduino board and a USB lead to connect the Arduino to your computer. For most projects, your Arduino board can be powered through the USB connector, either from your computer or from a power supply. The boards also can be powered using a direct-current (dc) adapter connected to the dc power socket on the same side as the USB socket.

The red button on both boards is the reset button. You will not need to press this much, if at all, with the Uno; however, if you use a Leonardo, you will need to press it sometimes when programming the board.

Installing Arduino

Before you can start making some of the projects in this book, you will need to set up your computer with the Arduino software so that you can program the Arduino from your laptop or desktop computer. The Arduino software is compatible with Windows, Mac, and Linux, although the installation instructions for each are different. For the latest up to date installation instructions for your platform, visit the official Arduino website (www.arduino.cc), where you will also be able to download the software and follow the installation instructions.

Making a Light-Emitting Diode (LED) Blink

Traditionally, the first thing that most books will teach you is how to make the built-in LED on the Arduino flash. This is a useful exercise for two reasons. First, it shows that everything is set up okay and that your computer can communicate with the Arduino to program it. Second, it is a nice simple example that uses the LED built onto the Arduino board, and therefore, you do not need anything except your Arduino, your computer, and a lead to connect the two.

Start the Arduino IDE software, and open the example "Blink" sketch (programs are called *sketches* in the Arduino world). You will find the sketch from the file menu, under "Examples" and then "Basic." With the sketch opened, you should see something like Figure I-3.

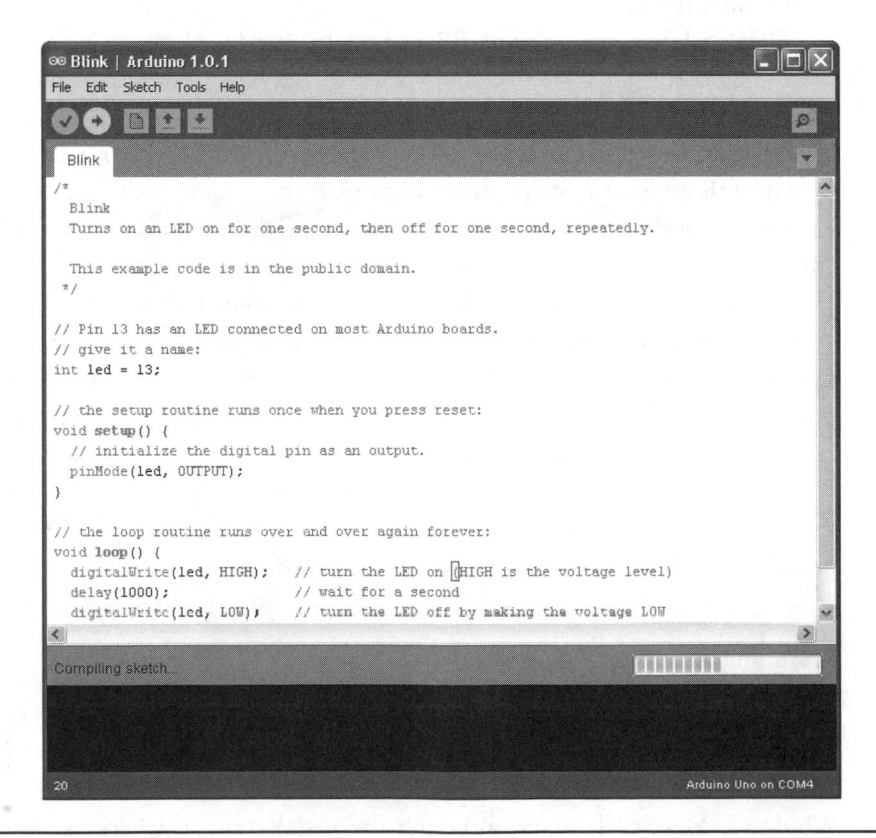

FIGURE I-3 Blink sketch.

Before you can program the Arduino board with this sketch, you need to set the board type and serial port from the "Tools" menu. If all is well, when you press the "Upload" button (highlighted in Figure I-3), there should then be some flashing of the LEDs on the board, and then the LED marked "L" on the board should start to blink slowly.

Next, try changing 1,000 to 200 in the two delay commands. Upload the sketch again, and the LED will blink at a much faster rate.

Protoshields

Many of the projects in this book make use of a Protoshield (Figure I-4). A *Protoshield* is a plain circuit board designed to sit over the top of an Arduino. It has a large area to which you can attach your own components. Although ready-made Protoshields are quite expensive to buy, you can also just buy the bare boards for a few dollars and attach your own header pins.

The first project in Chapter 1 uses a Protoshield to create a persistence-of-vision display that appears to paint a message in the air when you wave it from side to side. Figure I-5 shows this project. As you can see in the figure, eight LEDs,

FIGURE I-4 Protoshield.

FIGURE 1-5 Protoshield from Chapter 1 (persistence of vision).

eight resistors, and a small module (a tilt sensor) are attached to the board. The component leads are usually pushed through the holes in the top of the board and soldered to the pads underneath, and the leads of the components are soldered together, often with extra bridging wires, to make up the circuit. This first project explains in great detail exactly how to solder components to the Protoshield.

A number of different Protoshield designs are available on the market. The one used in this book is the official R3 Protoshield designed by the makers of Arduino. This is available from the Arduino store (http://store.arduino.cc/eu) for just €3 for a bare board. You will also find it for sale at many of Arduino's distributers and on eBay.

To be able to plug the Protoshield into your Arduino, you will also need some lengths of header pins. See the Appendix for more details about where to obtain components.

Figure I-6 shows the easiest way to make sure that the header pins are soldered on straight. First, break of lengths of 10, 8, 8, and 6 pins each, and push the long ends into your Arduino. Then place the shield over the top of the holes, and make sure that it is the right way up (Figure I-6*a*). The Arduino board will keep the pins straight while they are being soldered. Solder each pin in turn (Figure I-6*b*). When all the pins are soldered, the shield should look like Figure 6*c* when you turn it over.

a

b

Figure I-6 Soldering headers onto a Protoshield.

FIGURE I-6 Soldering headers onto a Protoshield (*continued*).

There are holes on the protoshield to add a reset switch, but this is not really necessary because the "Reset" button on the Arduino is still accessible even with the Protoshield fitted.

Components

The Appendix contains a list of all the components that you will need for the projects in this book, along with various sources. You will often find that many of the components can be obtained very cheaply on eBay.

The Book

The remainder of this book is organized into chapters that deal with a particular theme, but within that theme, there is no real order to the projects. The only project that gets something in the way of special treatment is the first project in Chapter 1. This project is to build a persistence-of-vision (POV) display. The chapter explains how to us a Prototshield and construct the project in more detail than for most other projects. Thus, if you are new to this type of construction, read Chapter 1 first, even if you do not plan to make the project.

The TAB Book of Arduino Projects

PART ONE

Light and Color

CHAPTER 1

Persistence-of-Vision Display

Difficulty: ★★	Cost guide: $15

Persistence-of-vision (POV) displays are readily available as novelty toys, but buying one ready-made is nowhere near as much fun as making one for yourself. Figure 1-1 shows the effect you get when you wave the display about vigorously. This POV display also has the advantage that you can program the message that it displays from your computer.

The project uses a Protoshield with seven light-emitting diodes (LEDs), series resistors, and a tilt-sensor module soldered to it.

Parts

To build this project, you will need the following:

Name	Quantity	Description	Appendix
	1	Protoshield printed circuit board (PCB) and header pins	A3, H1
	1	Sparkfun Tilt-a-Whirl tilt sensor	M1
R1–7	7	270 Ω, ¼ W resistors	R1
D1–7	7	5-mm LEDs (color of your choice)	S1–5

LEDs are often best bought from eBay, especially when you need them in quantity.

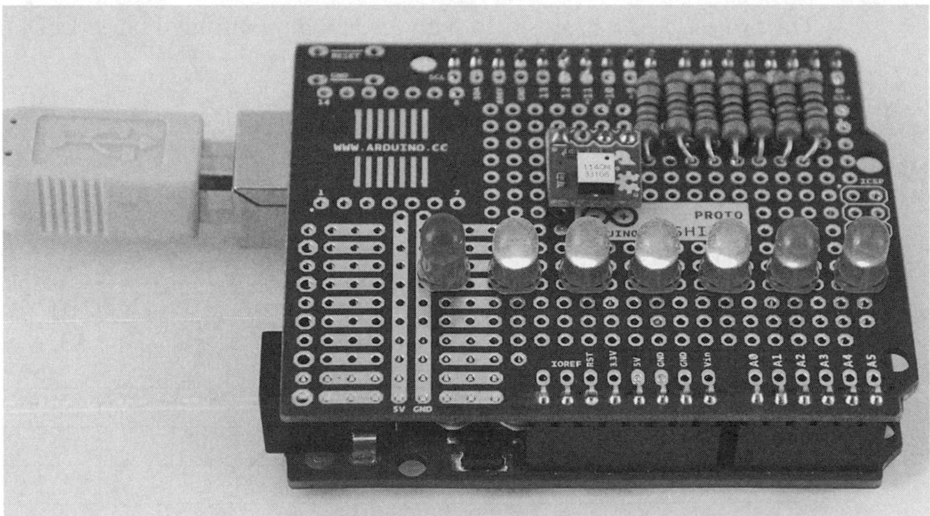

Figure 1-1 Persistence-of-vision display.

Protoshield Layout

Figure 1-2 shows the layout of the components on the Protoshield.

Construction

Because this is the first project in the book, I will go into some detail. The basic idea of Protoshield is that the component leads are pushed through from the top of the board and soldered to the pads beneath, and then the remaining leads of the components are joined up. Sometimes, as with all four leads from the tilt sensor, linking wires have to be soldered in place.

The general rule for soldering things onto any kind of circuit board is always to start with the lowest-lying components so that when you lay the board on its back, gravity will ensure that the components stay in position while you solder them.

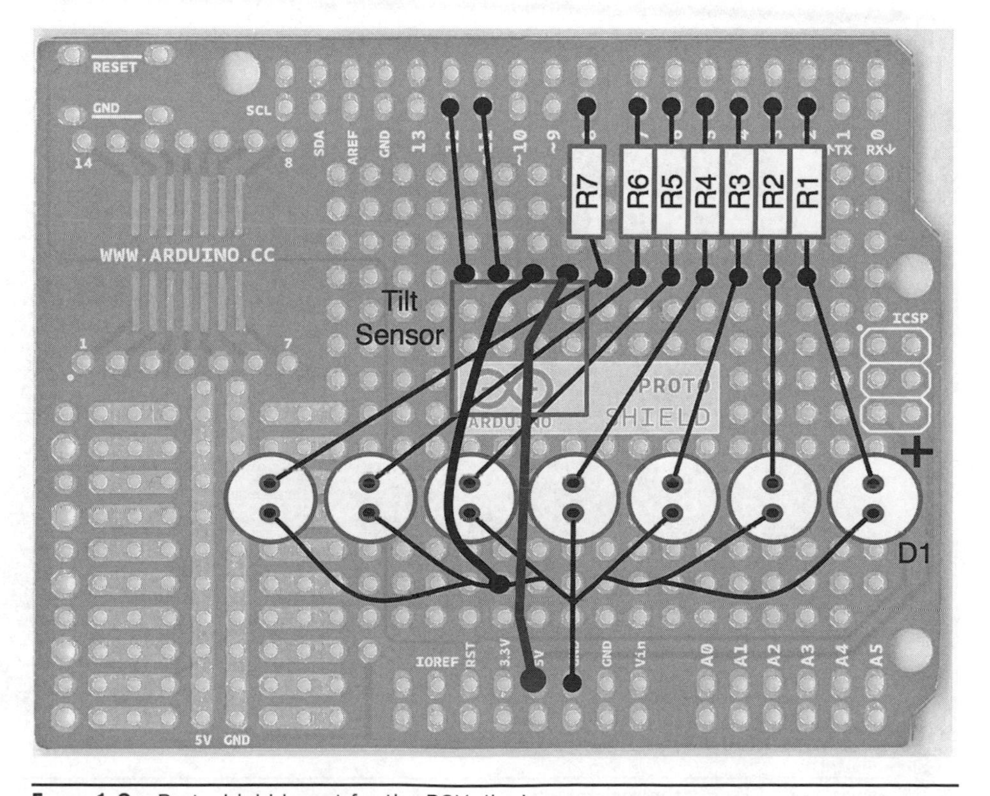

FIGURE 1-2 Protoshield layout for the POV display.

Step 1: Put the Resistors in Place

Using Figure 1-2 as a guide, bend the leads of the resistors, and push the leads through. Figure 1-3 shows the underside of the board with the resistors ready to be soldered into place.

Step 2: Solder the Resistors

Now solder the resistor leads next to the header pins, and cut off the excess leads on that side (Figure 1-4). Solder the pins on the other side of the resistors to the pads, but do not cut off the leads yet.

Step 3: Solder the LEDs to the Resistors

Now you can start soldering the LEDs. LEDs have a positive end and a negative end, so you need to make sure that the longer positive leads are all toward the

Figure 1-3 Placing the resistors.

Figure 1-4 Soldering the resistor leads.

resistors. All the negative leads of the LEDs are eventually going to be connected together.

Figure 1-5 shows the first LED in place. Bend the positive lead of the LED and the free lead of the first resistor so that they run right next to each other (the far right of Figure 1-5), cut off the excess lead, and solder them together.

When all of the LEDs have been soldered to their corresponding resistor leads, the underside of the board should look like Figure 1-6.

Step 4: Solder the LED Negative Leads

The negative leads of the LEDs all need to be connected together and then to the Arduino ground (GND). Start by shortening the negative lead of the middle LED so that it just reaches the pad for the GND pin; then bend the other leads in until they are all connected together (Figure 1-7).

FIGURE 1-5 Soldering the LEDs—1.

FIGURE 1-6 Soldering the LEDs—2.

FIGURE 1-7 Soldering the LEDs—3.

Step 5: Test the LEDs

When you flip the board over, you can see that you have a neat little board of LEDs (Figure 1-8). We can actually test this before we go any further.

Plug the shield on top of the Arduino, and upload sketch ch_01_pov_led_test onto the Arduino. The LEDs will each light in turn. All the programs (or sketches as they are called in the Arduino world) are available as a download from the author's website at www.simonmonk.org. Just follow the link for this book, and then look for the "Code" link.

If any of the LEDs do not light, then check that they are wired up the right way and that there are no accidental solder connections.

Step 6: Prepare the Tilt Sensor

The tilt sensor is used to detect when the POV display should start flashing the LEDs to make the message appear in the air. The tilt sensor used in this project is a module sold by SparkFun. It comes as a tiny circuit board, so we need to attach some header pins to it to link it to the Protoshield. Figure 1-9 shows the sensor with the header pins attached.

Figure 1-8 Testing the LEDs.

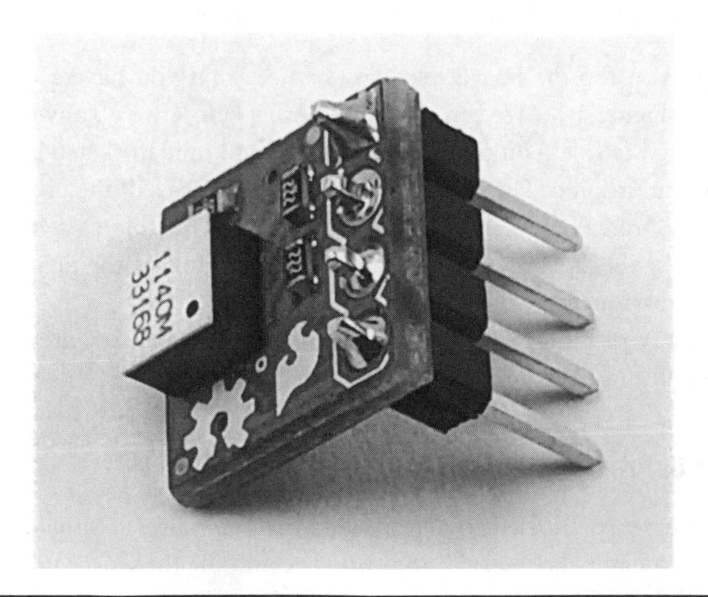

Figure 1-9 Preparing the tilt sensor.

The tilt sensor has four connections: GND, 5-V, and two switch connections that detect tilting in two axes.

Step 7: Solder the Tilt Sensor

Unlike the LEDs and resistors, the tilt sensor does not have long leads that we can use for connecting things up. Instead, use some of the trimmed resistor leads to connect the solder pads for Arduino pins 11 and 12 to the appropriate pins of the tilt sensor (refer to the wiring diagram of Figure 1-2). The end result is shown in Figure 1-10.

The 5 V and GND connections to the tilt sensor need to be made with insulated wire because they cross over other connections. These are shown in Figure 1-11.

Testing

Now that construction is complete, we can try out the full project. Program your Arduino with the sketch ch_01_pov and then attach the Protoshield. All the

FIGURE 1-10 Soldering the tilt sensor—1.

Figure 1-11 Soldering the tilt sensor—2.

programs (or sketches as they are called in the Arduino world) are available as a download from the author's website at www.simonmonk.org

Now, if you wave the Arduino and Protoshield back and forth, you should see the word *Arduino*. You can change the text that is displayed by opening the Arduino serial monitor using the rightmost icon on the toolbar. Type a different message at the top, and click "Send." The message will be transferred to the LEDs.

Software

The code responsible for actually flashing the LEDs is fairly concise, but there is also a big array in the sketch that contains the character font in a seven-row format. The first part of the sketch is as follows:

```
// Project 1. POV


const int MAX_MESSAGE_LEN = 40;
const int gap = 700;
```

```
const int sw1Pin = 11;
const int ledPins[] = {2, 3, 4, 5, 6, 7, 8};

char message[MAX_MESSAGE_LEN];

extern byte characters[96][7]; // so the array can be defined
                               // at the end
```

If you need to flash longer messages than 40 characters, then you can change the value of `MAX_MESSAGE_LEN`. The constant *gap* sets the delay in microseconds between the display of each column of the text. You can lower this value to make the message more condensed. Increasing the value will spread the message out wider.

The following setup function initializes the pin attached to the tilt sensor to be an input and all the LED pins to be outputs. It also uses the C `strcpy` command to copy the message to be displayed into the buffer message and initializes serial communication with your computer.

```
void setup()
{
  pinMode(sw1Pin, INPUT_PULLUP);
  for (int i = 0; i < 7; i++)
  {
    pinMode(ledPins[i], OUTPUT);
  }
  strcpy(message, "Arduino"); // default message
  Serial.begin(9600);
}
```

The following `loop` function first checks to see if a new message string has been sent from the computer. If it has, it copies it into the `message` buffer. It then goes on to check the tilt switch. If movement is detected, it first delays for 10 columns worth of time, to allow the display to get up to speed, and then loops over each letter in the message calling the function `displayChar` to actually flash out the columns for that particular letter.

```
void loop()
{
  if (Serial.available())
  {
```

```
    int len = Serial.readBytesUntil(0, message, MAX_MESSAGE_LEN);
    message[len] = 0; // string terminator
  }
  int n = strlen(message);
  if (digitalRead(sw1Pin) == LOW)
  {
    delayMicroseconds(gap * 10);
    // play it!
    for (int i = n-1; i >= 0; i--)
    {
      displayChar(message[i]);
    }
  }
  delayMicroseconds(gap * n * 9);
}
```

To flash out a particular character, the `displayChar` function first checks that the ASCII value of the character is in the range of 32 to 127 (decimal). ASCII (American Standard Code for Information Interchange) assigns a number to every character, for example, `A` is 65, `B` is 66, and so on.

The function then has a pair of nested loops—the outer loop for each column and the inner loop for each row of a particular column. The `characters` array is used to look up the bit for that row and column to see if the appropriate LED should be lit. At the end of each character, a delay of twice the normal column gap allows some separation between the letters of the message.

```
inline void displayChar(char ch)
{
  if (ch >= 32 && ch <= 127)
  {
    for (int column = 0; column < 8; column++)
    {
      for (int row = 0; row < 7; row++)
      {
        byte rowByte = characters[ch - 32][row] >> 1;
        int pixel = bitRead(rowByte, column);
        digitalWrite(ledPins[row], pixel);
      }
      delayMicroseconds(gap);
    }
```

```
  }
  delayMicroseconds(gap * 2);
}
```

The LED pins used are kept in an array `ledPins`. This allows them to be iterated in a loop when setting them to be outputs.

The array element for the letter `A` is shown below using binary numbers. A `1` means that the LED will be on; a `0` means that it will be off. You can just make out the pattern of the `A`.

```
  {
    //65 A
    0b00100000,
    0b01010000,
    0b10001000,
    0b11111000,
    0b10001000,
    0b10001000,
    0b00000000
  },
```

Summary

This is a fun little project to start with. In the next project, we will continue with the theme of LEDs.

CHAPTER 2

LED Cube

Difficulty: ★★★★	Cost guide: $10

LED cubes require some care and good soldering skills to make, but they make very effective light displays. Every one of the 27 LEDs in this project can be turned on and off independently, and you can make some really interesting animation effects even with just 3 × 3 × 3 pixels (Figure 2-1).

For some truly amazing LED cubes, with thousands of LEDs, search YouTube. Be warned that even a 3 × 3 × 3 cube is quite fiddly to make.

Parts

To build this project, you will need the following:

Name	Quantity	Description	Appendix
	1	Protoshield printed circuit board (PCB) and header pins	A3, H1
R1–9	9	270 Ω, ¼ W resistors	R1
D1–27	27	5-mm LEDs (color of your choice)	S1–5
T1–3	2	2N7000 MOSFET transistors	S6
		Solid-core wire	
		Cardboard, scissors, a drill and 5-mm drill bit	

FIGURE 2-1 A 3 × 3 × 3 LED cube.

LEDs are often best bought from eBay, especially when you need them in quantity. In this project, look for LEDs with as wide an angle of view as possible so that not all the light is shining straight up. Otherwise, the lower LEDs just illuminate the LEDs on the row above, reducing the effect of the display.

Protoshield Layout

Figure 2-2 shows the layout of the components on the Protoshield. The resistors and transistors are soldered to the Protoshield, before any of the LEDs are attached. The LEDs will be built up in layers onto the board. The points marked

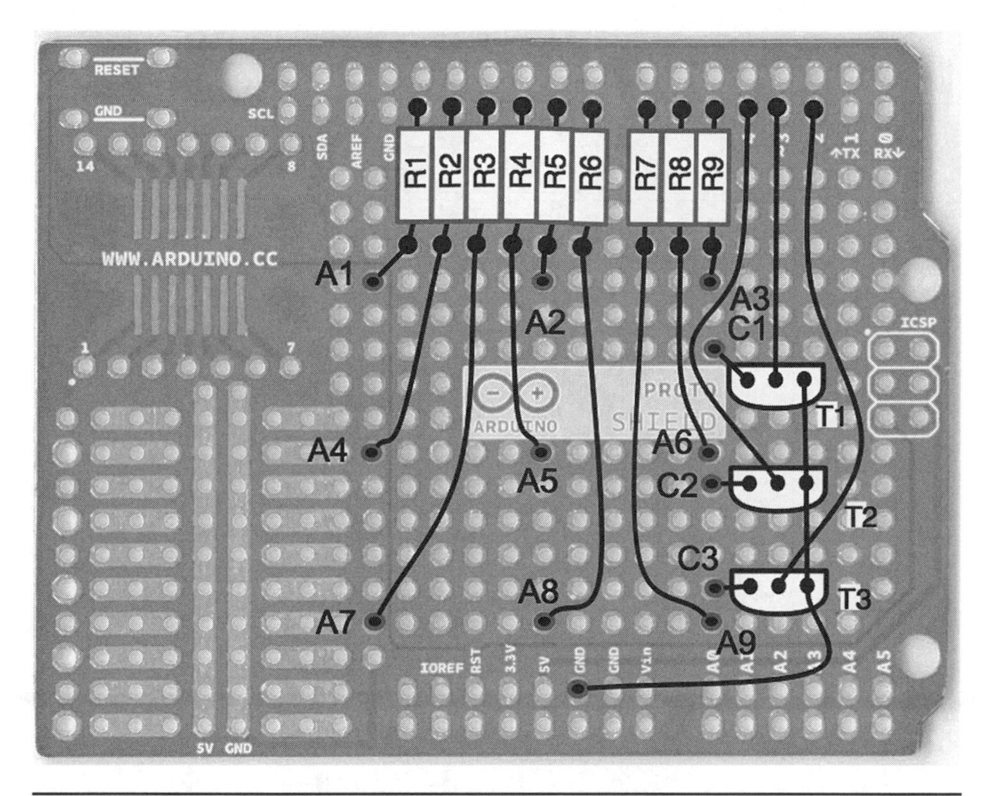

FIGURE 2-2 Protoshield layout for the LED cube.

A1 to A9 and T1 to T3 are the points where the LED assembly eventually will be connected to the Protoshield.

Figure 2-3 shows the schematic diagram for the design. You do not have to understand it to build the design, but it may help to explain why everything is connected up the way it is. Each layer of the cube is made up of nine LEDs in a 3 × 3 grid. The anodes (positive connections) of each LED position are connected together for all three layers. Thus, one Arduino pin could turn on all three LEDs in any give column of LEDs. However, the three transistors, T1 to T3, can switch each of the layers on and off. The Arduino sketch will rapidly turn each layer on in turn, setting the appropriate LEDs on for that layer, before moving on to the next.

Construction

There are two major parts to building this project. The first is to assemble the Protoshield, and the second is to create the LED assembly itself.

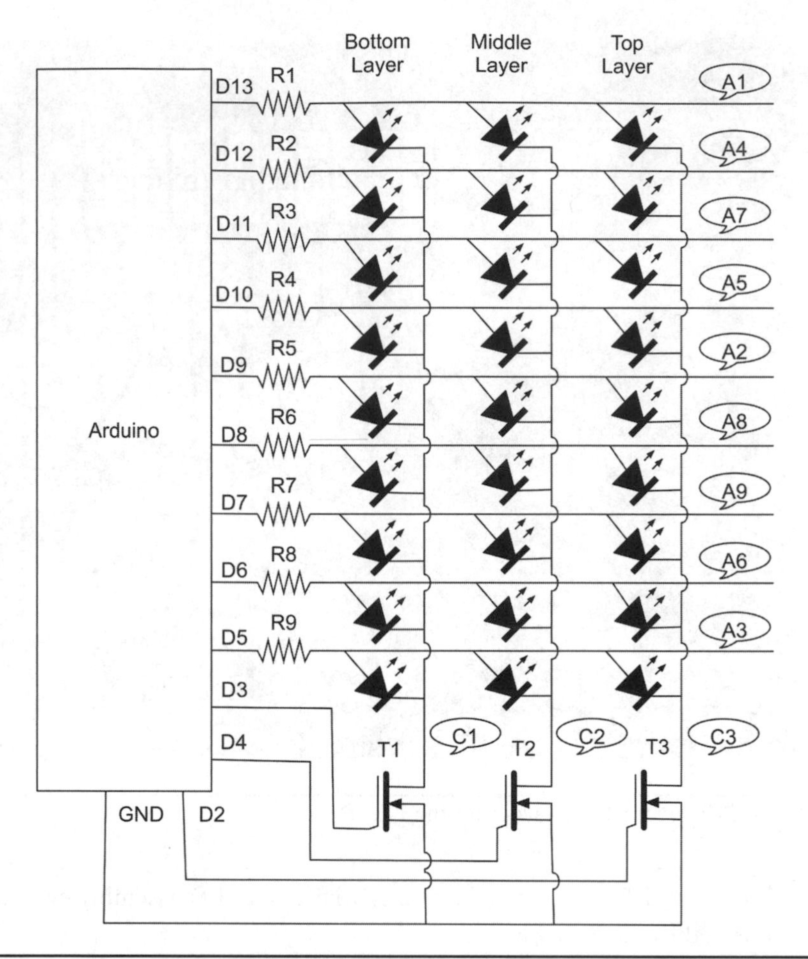

Figure 2-3 Schematic diagram for LED cube.

The starting point is a Protoshield with the header pins attached. If you are not clear how to do this, please return to Chapter 1.

Step 1: Solder the Resistors

Using Figure 2-2 as a guide, bend the leads of the resistors, and push the leads through. Solder the resistors in place, and cut off the excess leads at the edge near the pin headers. However, leave the resistor leads toward the middle of the board in place because you will need these to make connections to the LEDs later. Figures 2-4 and 2-5 show how things should look on the bottom and top of the board when you have done this.

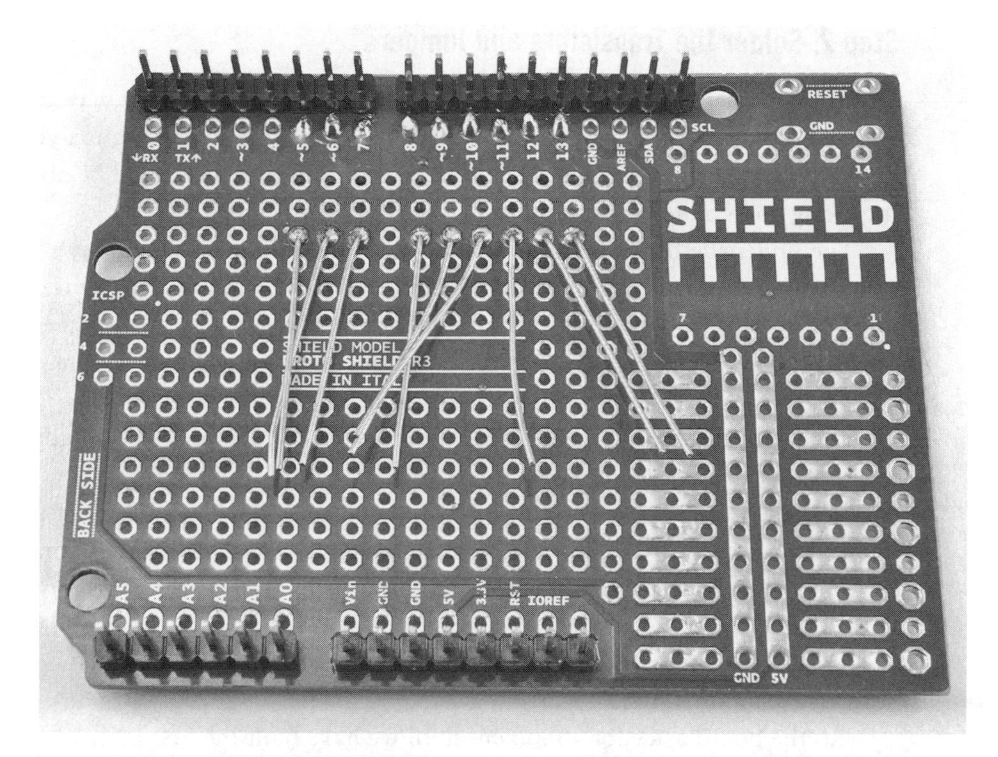

FIGURE 2-4 Soldering the resistors.

FIGURE 2-5 The resistors in place.

Step 2: Solder the Transistors and Jumpers

Now you can solder the transistors. Make sure that they are the right way around; you can see the transistors have one flat side in Figure 2-6. In this figure, you can also see the bare wire coming vertically out of the board. This should be about an inch long and connects to the point marked "C1" in Figure 2-2.

The underside of the board is shown in Figure 2-7, and you can see that the other connections for the transistors have been made with insulated wire or by bending the leads of the transistors and soldering them together to form connections.

Use Figure 2-2 as a guide to making these connections. They are

- The right-hand three leads of the transistors all connected together and connected to GND
- The middle connection of T1 connected to Arduino pin D3, the middle connection of T2 connected to Arduino pin D4, and the middle connection of T3 connected to Arduino pin D2

The resistor leads are still unconnected. These will not be connected until we start to attach the LEDs.

Put the board aside for a moment until we have built the first layer of LEDs.

Figure 2-6 Soldering the transistors.

FIGURE 2-7 Soldering the transistors (bottom).

Step 3: Prepare a Holder for the LEDs

When connecting the LEDs together, you are going to solder the LED leads directly to each other. This is almost impossible without some kind of mechanism for keeping all the LEDs in the right place while you solder them.

For this task, I made a simple holder out of cardboard. You could, if you wish, use something stiffer, such as wood, but cardboard will do the job. Figure 2-8 shows the cardboard holder. To make it, cut out a piece of thick cardboard about 2 inches (50 mm) square. Next, draw a grid of lines on it ½ inch (12 mm) apart. The holes are going to be drilled at the nine intersections of these lines.

Drilling may seem a bit excessive for cardboard, but drilling a hole with just the right diameter for the LEDs (5 mm) keeps the LEDs snugly in place while you are trying to solder them. Figure 2-9 shows the pattern of the LEDs. This is the same for all three layers of the board.

Each LED has a positive lead (the longer lead). This lead will always be left unbent. However, all the negative leads of the LEDs are going to be connected together by bending the leads to a horizontal position so that they just reach to the next LED.

FIGURE 2-8 LED holder.

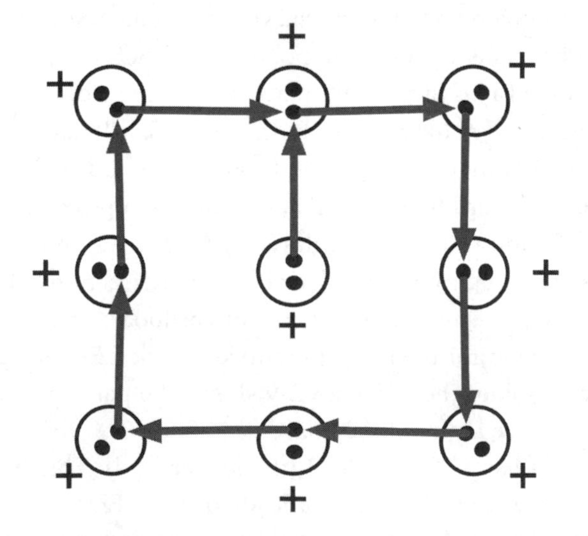

FIGURE 2-9 LED connections.

Place the LEDs into the holder with their leads facing upward. Note that with the exception of the middle LED, all the LEDs have their longer positive lead toward the outside of the square.

Step 4: Make the First Layer of LEDs

Start in the top left corner, and bend the short negative leads of the LEDs in the directions indicated in Figure 2-9. Note that the arrow marks the endpoint of the lead. Figure 2-10 shows the first row of LEDs soldered up. Remember to snip off any excess lead after soldering.

When all the LEDs are in place, the holder should look like Figure 2-11, and you can now remove the LEDs from the holder (Figure 2-12). Be careful with the assembly; once the LEDs are bent out of position, the assembly will never look as neat.

Figure 2-10 The LEDs in the cardboard holder.

FIGURE 2-11 The LEDs all connected.

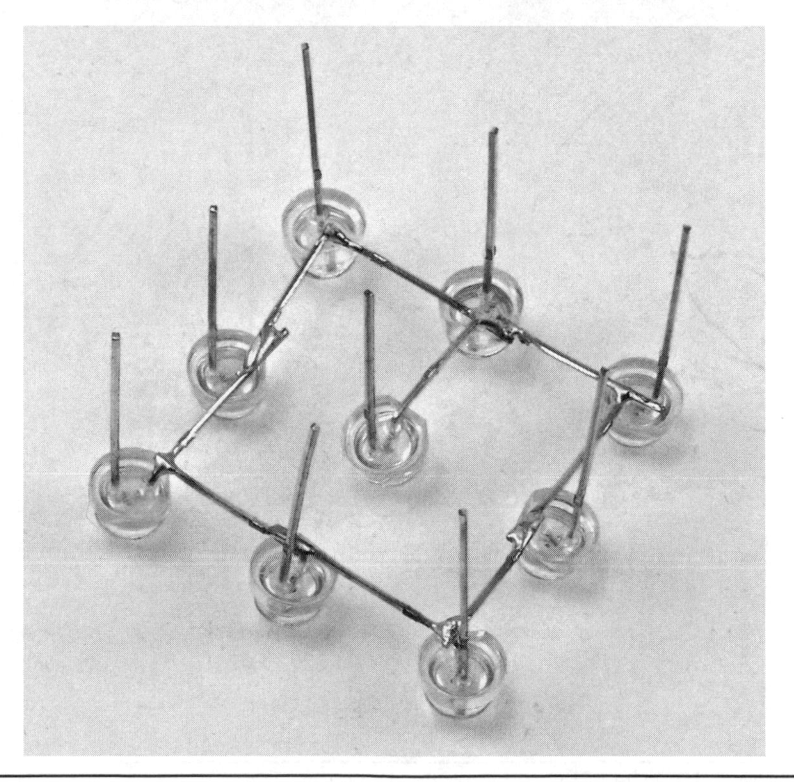

FIGURE 2-12 The bottom layer of LEDs.

Step 5: Attach the Bottom Layer of LEDs

It is now time to fix the first layer of LEDs onto the Protoshield. Carefully edge each of the vertical LED pins into the correct hole in the Protoshield. Refer to Figure 2-2. Push them all through by the same amount, which should be just enough to solder to but no longer.

It is easiest to do this by starting at one edge of the LEDs and working across. Placing the board over some kind of jam jar or pot will prevent the LEDs from getting squashed out of shape (Figure 2-13).

After the LED pins are soldered, the resistor leads that you have not yet soldered should be soldered to the LED connections. Finally, the vertical lead coming straight out for the board from the transistor T1 should be soldered to the common negative connections of the LEDs. Make sure that the lead is not touching anything, and snip off any excess.

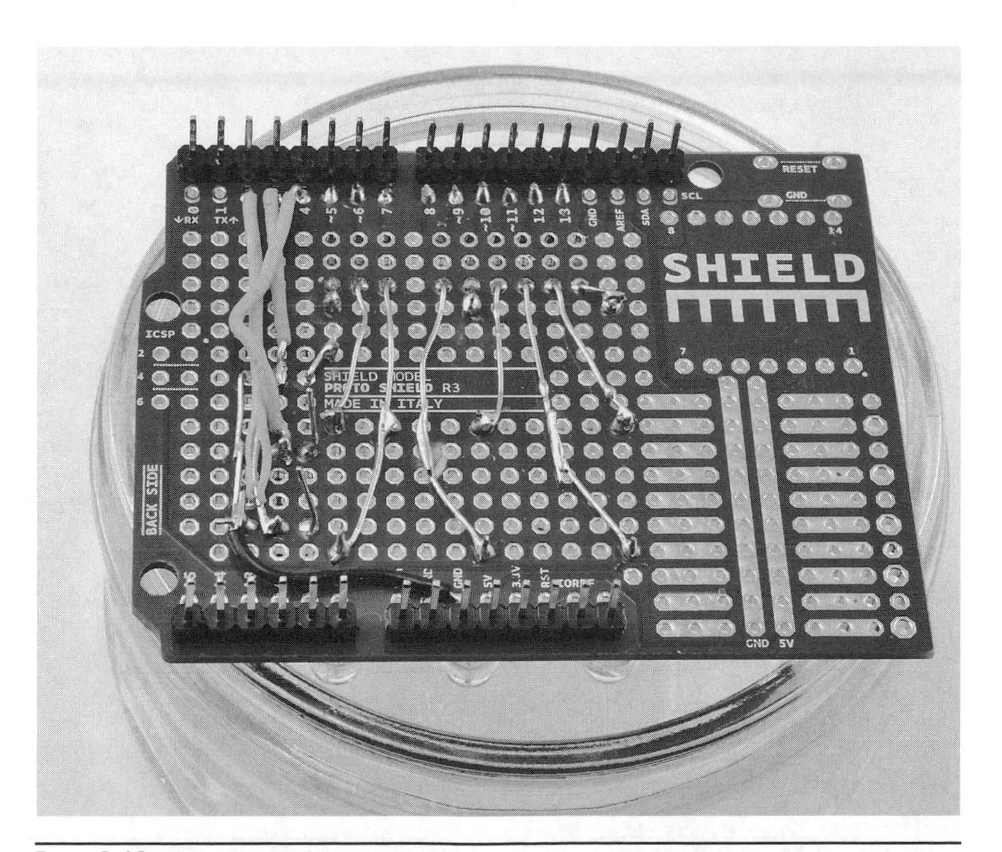

Figure 2-13 Attaching the bottom layer of LEDs.

Step 6: The Second Layer

Make the second layer of LEDs in exactly the same way as you did the first. The only difference is that when you have finished, you will need to bend over the tips of the vertical positive leads (Figure 2-14) so that they can reach around the LEDs on the first layer and connect to the corresponding positive leads on that layer (Figure 2-15). The leads are best bent using pliers.

Soldering the second layer onto the first is a little tricky, especially reaching the middle LED. Having a "helping hands" stand helps a lot because you can tilt the board to just the right angle.

Start by soldering the positive leads of each corner LED, and then solder the other leads. Finally, do not forget to attach a vertical wire from the common negative of the middle layer down to connection point C2 on the Protoshield.

FIGURE 2-14 Bending leads on the second layer.

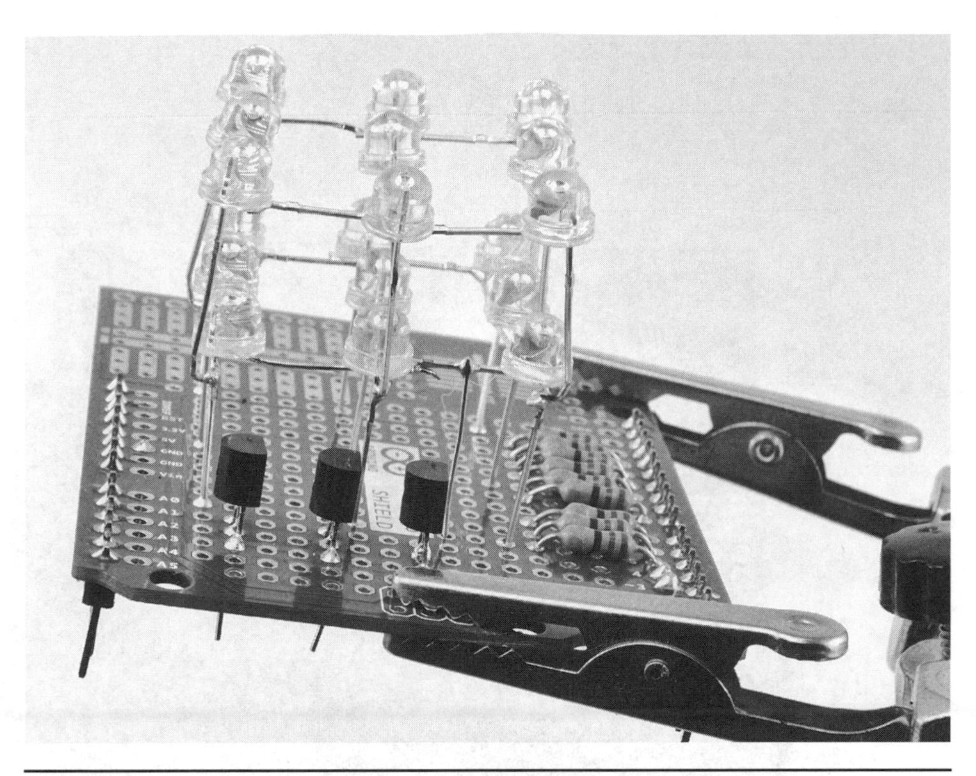

Figure 2-15 Soldering the second layer to the first.

Step 7: The Top Layer

Build the top layer exactly the same as you did the second layer, and attach it to the second layer in the same way. Do not forget to attach a vertical linking wire to connect the common negative of the top layer to connection point C3 of the Protoshield.

Figure 2-16 shows the final shield fully assembled. Note the connecting wires to the common negatives of each layer.

Using the LED Cube

Before you plug the shield onto your Arduino, make a very careful visual inspection of the LED module. Check that none of the LEDs has its pins accidentally connected together.

When you are sure that everything is okay, then load sketch ch_02_led_cube onto your Arduino, and attach the shield. All the programs (or sketches as they are

Figure 2-16 LED cube fully assembled.

called in the Arduino world) are available as a download from the author's website at www.simonmonk.org.

You should see the LEDs twinkle on and off at random. There are some interesting other visual effect sketches that can be downloaded from the website; all the sketch names begin with ch_02.

Software

The example sketch for this hardware uses a two-dimensional array to contain the LED pins (`ledPins`). It also uses an array for the `layerSelect` pins that control which layer is enabled during refreshing of the display.

A separate three-dimensional array (`model`) is used to maintain the current state of the 27 LEDs. By setting any one of its locations `HIGH`, the corresponding LED will be lit during refresh.

```
void refresh()
{
  for (int layer = 0; layer < 3; layer++)
  {
    digitalWrite(layerSelectPins[layer], HIGH);
    for (int x = 0; x < 3; x++)
    {
      for (int y = 0; y < 3; y++)
      {
        digitalWrite(ledPins[y][x], model[layer][y][x]);
      }
    }
    delay(refreshPeriod);
    digitalWrite(layerSelectPins[layer], LOW);
  }
}
```

The `refresh` function should be called as frequently as possible from the main loop. It enables each layer in turn and sets the LEDs according to the `model` array.

```
void loop()
{
  static long lastUpdate;
  long now = millis();
  if (now > lastUpdate + period)
  {
   animate();
   lastUpdate = now;
  }
  refresh();
}

void animate()
{
   model[random(3)][random(3)][random(3)] = HIGH;
   model[random(3)][random(3)][random(3)] = LOW;
}
```

The animated flickering is caused by the `animate` function turning one LED on and one LED off at random each time it is called.

Summary

This is a great little novelty project, but one that will have tested your construction skills. The next project also uses LEDs, but this time high-power LED panels.

CHAPTER 3

High-Power LED Controller

Difficulty: ★★	Cost guide: $20

This project uses an Arduino to control the power to 12 V LED modules using high-power MOSFET transistors. You can use pretty much any 12 V direct-current (dc) lights with this project that are marked as being "dimmable." You can also expand it to control six rather than just the three channels I used (Figure 3-1).

FIGURE 3-1 High-power LED lighting controller.

Parts

To build this project, you will need the following:

Name	Quantity	Description	Appendix
	1	Protoshield printed circuit board (PCB) and header pins	A3, H1
T1–6	3 or 6	FQP30N06 MOSFETs	S7
R1–6	3 or 6	270 Ω, ¼ W resistors	R2
D1–6	3 or 6	12 V LED module	M2
	1	12 V, 1 to 2 A power supply	M3
	1	Strip of 0.1-inch header socket (six-way)	H2

The MOSFETs have a maximum current rating of 30 A. However, the wiring through the Arduino PCB should not be pushed further than, say, 1 or up to 3 A as long as the LEDs are not at full brightness all the time. Thus, quite high-power LED modules of up to, say, 5 W should be fine.

When selecting your power supply, make sure that it can supply enough power for all the LEDs to be on at the same time. Thus, if there are six lights each using 3 W, total power is 6 × 3 = 18 W, and the total current is 18 W/12 V = 1.5 A.

Protoshield Layout

Figure 3-2 shows the schematic diagram for the project, and Figure 3-3 shows the Protoshield layout. The schematic just shows three channels. Adding more channels just means adding another MOSFET and resistor for each extra channel. The Protoshield layout shows the arrangement for all six channels. If you just plan to build the three-channel version, then ignore everything inside the dashed box.

Construction

This is a pretty straightforward project to build. As always, start by attaching header pins to the Protoshield. These instructions assume that you are building the three-channel version of the project.

Step 1: Solder the Resistors

Using Figure 3-3 as a guide, solder resistors R1 through R3 in place. Cut off the excess leads at the Arduino connector end of the resistors, but leave the ends toward the middle of the board for making connections to the MOSFETs (Figures 3-4 and 3-5).

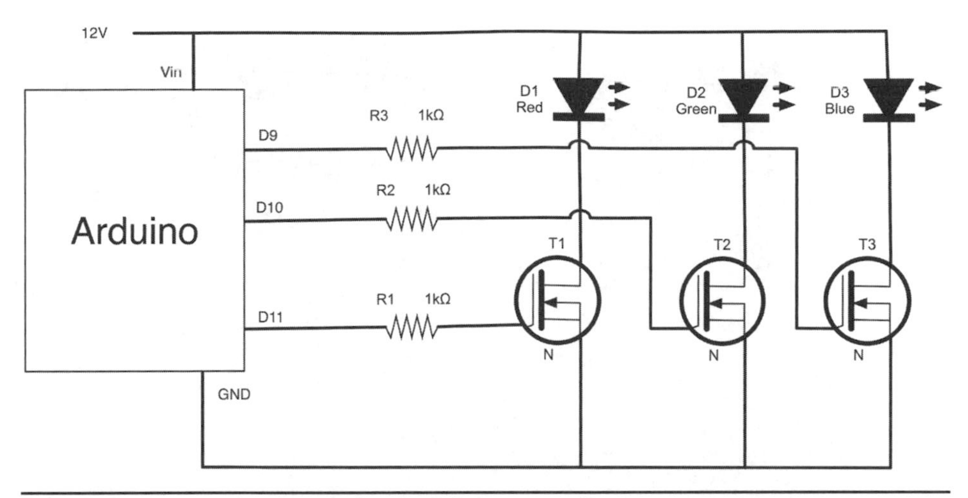

FIGURE 3-2 Schematic diagram for the lighting controller.

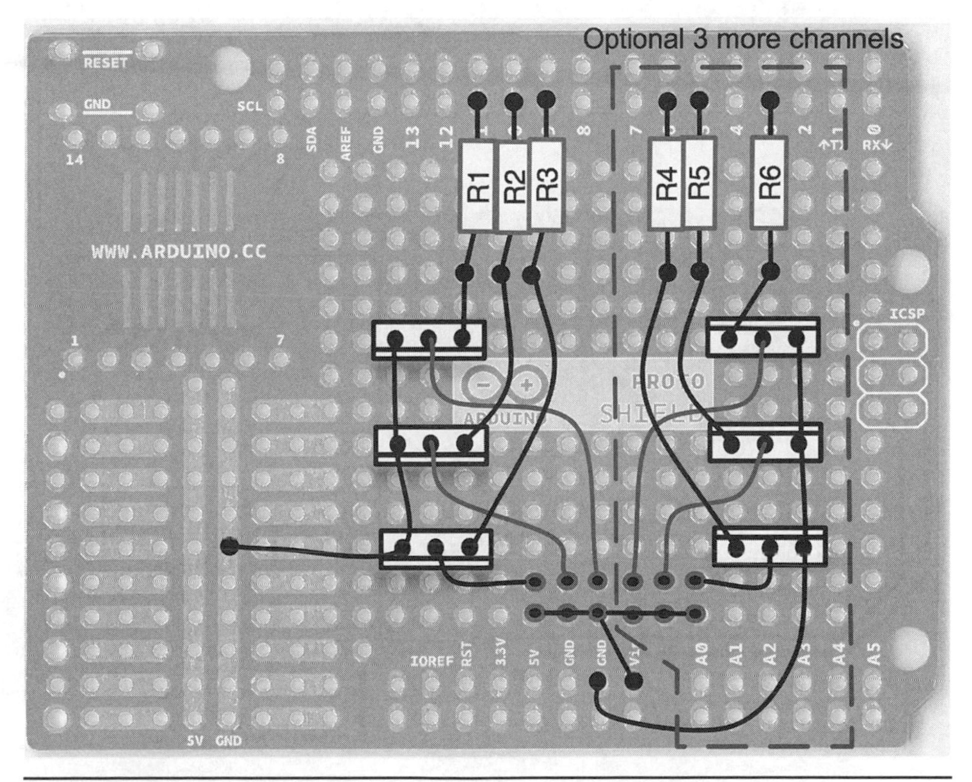

FIGURE 3-3 Protoshield layout for the lighting controller.

Figure 3-4 Soldering the resistors.

Figure 3-5 Resistors in place.

Step 2: Prepare the Header Sockets

Header sockets do not usually come in the right lengths. Typically, you will buy a strip of maybe 40 header sockets that need cutting to the right length.

Cutting the header sockets can be a little tricky (Figures 3-6 and 3-7). The sockets are very close together, so you will need to cut through the seventh socket, sacrificing one socket. You can cut them with a craft knife or just score a line with the craft knife on the seventh socket, and then break it over the edge of your desk or table.

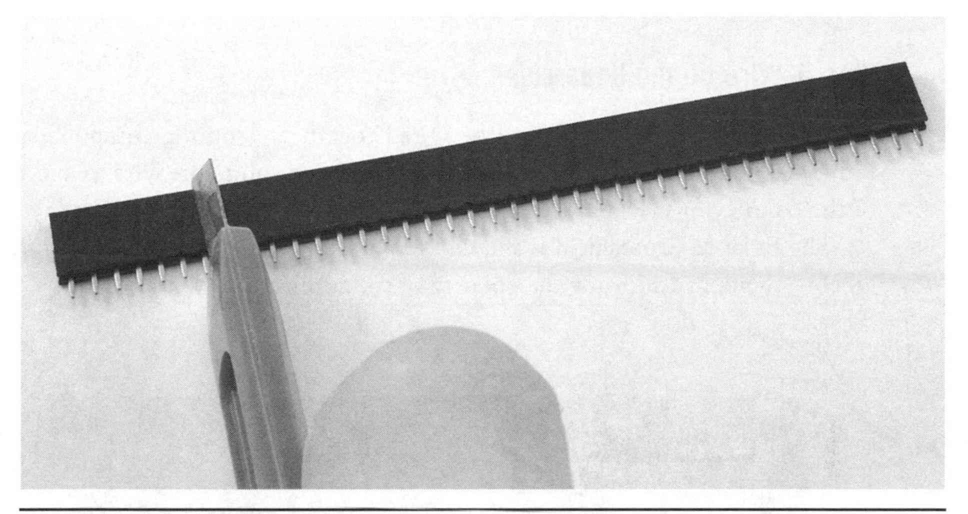

FIGURE 3-6 Cutting header-socket strips.

FIGURE 3-7 Two times six-way headers.

Step 3: Solder the Header Sockets

Although we are only going to use three channels, it is probably a good idea to solder all six socket pairs. It can be difficult to get the header sockets soldered straight (Figure 3-8) and to avoid having them fall out at the crucial moment. A bit of adhesive putty will hold them in place while they are being soldered.

Step 4: Solder the MOSFETs

When soldering the MOSFETs, make sure that they are the right way around, as shown in Figure 3-9.

Step 5: Wire Up the Underside

Quite a lot of the underside can be wired together using the component leads. However, you will require some additional insulated solid-core wire to make some of the connections (Figures 3-10 and 3-11). Refer to Figure 3-3 for the layout.

The finished Protoshield is shown in Figure 3-12. It is a good idea to attach a label to the board showing the polarity of the connections to the LEDs.

Figure 3-8 Soldering the header sockets.

Figure 3-9 Soldering the MOSFETs.

Figure 3-10 Wiring using component leads.

FIGURE 3-11 Final wiring.

FIGURE 3-12 The fully assembled Protoshield.

The LED modules that I used are panels made up of 36 LEDs in three colors: red, green, and blue. These have leads already attached that terminate in pin headers. If you would prefer screw terminals, then you will notice that some of the holes on the right-hand edge of the Protoshield are larger than the others. This is so that up to five 0.2-inch PCB screw terminals can be attached. You could fit four screw terminals, one being the common +12-V connection and the other three for three channels. The light panels that I used have self-adhesive tabs on the back, which were attached to the side of a project box and the wires threaded inside.

Using the LED Lighting Controller

This is one project that cannot be powered from the USB connection of your computer. This is so partly because the USB connection provides only 5 V and we need 12 V but also because the USB connection supplies only about 500 mA and we need more than that for the LEDs. Thus, in this case, an external power adaptor is used to supply the Arduino and shield.

Upload the sketch ch_02_lighting_controller to your Arduino, put the shield in position, and connect the power adaptor. You should find that each channel cycles around in brightness. Note that you can now unplug the USB lead from the Arduino to your computer because it is only needed for programming.

Software

There are lots of interesting control programs that you could write for your lighting controller. This example just cycles the brightness of each channel.

```
// Chapter 3. Lighting Controller
int redPin = 11;
int greenPin = 10;
int bluePin = 9;

int red = 0;
int green = 85;
int blue = 170;

void setup()
{
  pinMode(redPin, OUTPUT);
```

```
  pinMode(greenPin, OUTPUT);
  pinMode(bluePin, OUTPUT);
}

void loop()
{
  red ++; if (red > 255) red = 0;
  analogWrite(redPin, red);
  green ++; if (green > 255) green = 0;
  analogWrite(greenPin, green);
  blue ++; if (blue > 255) blue = 0;
  analogWrite(bluePin, blue);
  delay(20);
}
```

Something else that you could do with the code would be to control the device through the serial monitor, sending commands that then change the brightness of each color channel to allow a color to be mixed.

Summary

In this chapter, we have made a high-power LED lighting controller. In Chapter 4 we turn our attention to sensing light rather than controlling it.

CHAPTER 4

Color Recognizer

Difficulty: ★	Cost guide: $10

This is a really simple project. In fact, it is so simple that it does not even need a Protoshield board. The color-sensing module can just be attached directly to the Arduino (Figure 4-1). When the module is placed next to an object, it will determine its color.

Parts

To build this project, you will need the following:

Quantity	Description	Appendix
1	Color sensor module	M7
3	Male-to-female jumper wires	H6

Really, that's it! The only tricky bit might be tracking down the module. Modules normally can be found on eBay or at DealExtreme. Look for a module that uses a TCS3200. There are several different variations on this chip, but they all work in the same way. The chip has a transparent case, and dotted over its surface are photodiodes with different color filters over them (red, green, and blue). You can read the relative amounts of each primary color.

Design

The integrated circuit (IC) does not produce an analog output but instead varies the frequency of a train of pulses. You choose which color the pulse frequency corresponds to by changing the values on the digital inputs S2 and S3. Table 4-1 lists the pins on this module.

Construction

The module will fit directly into the Arduino (Figure 4-1) facing outward. You will need to make the following connections:

FIGURE 4-1 Color-sensing module attached to an Arduino.

- S0 module to D3 Arduino
- S1 module to D4 Arduino
- S2 module to D5 Arduino
- S3 module to D6 Arduino
- OUT module to D7 Arduino

You will also need three male-to-female jumper leads to connect

- VCC module to 5V Arduino
- GND module to GND Arduino
- OE module to GND Arduino

TABLE 4-1 Color-Sensing Module Pinout

Pin	Description	Description	Pin
S0	S0 and S1 select the frequency range. Both should be set HIGH.	2.5 to 5.5 V	VCC
S1		Ground	GND
S2	Red: S2 and S3 LOW. Green: S2 and S3 HIGH. Blue: S2 LOW, S3 HIGH. White: S2 HIGH, S3 LOW.	Output enable: Set to LOW to effectively turn the chip on.	OE
S3		Tie to ground with the attached jumper to turn the LEDs on.	LED
OUT	The output pulses	—	GND

Software

The sketch ch_04_color_sensing demonstrates the use of this module.

```
// color_sensing
int pulsePin = 7;
int prescale0Pin = 3;
int prescale1Pin = 4;
int colorSelect0pin = 5;
int colorSelect1pin = 6;
```

The pins are named according to their function rather than using the module pin names. The `setup` function sets the appropriate pin modes and then sets the `prescale` pins that control the output frequency range to `HIGH`, starts serial communication, and then displays a welcome message.

```
void setup()
{
  pinMode(prescale0Pin, OUTPUT);
  pinMode(prescale1Pin, OUTPUT);
  // set maximum prescale
  digitalWrite(prescale0Pin, HIGH);
  digitalWrite(prescale1Pin, HIGH);
  pinMode(colorSelect0pin, OUTPUT);
  pinMode(colorSelect1pin, OUTPUT);
  pinMode(pulsePin, INPUT);
  Serial.begin(9600);
  Serial.println("Color Reader");
}
```

The `loop` function reads the three different colors (more on this later) and displays a message depending on the dominant color. Note that the lower the value, the brighter is that particular color.

```
void loop()
{
  long red = readRed();
  long green = readGreen();
  long blue = readBlue();
  if (red < green && red < blue)
  {
    Serial.println("RED");
  }
  if (green < red && green < blue)
  {
    Serial.println("GREEN");
  }
  if (blue < green && blue < red)
  {
    Serial.println("BLUE");
  }
  delay(500);
}
```

Each of the functions `readRed`, `readGreen`, `readBlue`, and `readWhite` just call a function `readColor` with the appropriate values for S2 and S3.

```
long readRed()
{
  return (readColor(LOW, LOW));
}
```

The function `readColor` first sets the appropriate pins for the color and then records a start time in the variable `start`. Then it waits for 1,000 pulses to happen and returns the difference between the current and start times.

```
long readColor(int bit0, int bit1)
{
  digitalWrite(colorSelect0pin, bit0);
  digitalWrite(colorSelect1pin, bit1);
  long start = millis();
  for (int i=0; i< 1000; i++)
  {
    pulseIn(pulsePin, HIGH);
  }
  return (millis() - start);
}
```

Although not actually used, there is also a function in the sketch that writes the color values to the serial monitor.

```
void printRGB()
{
  Serial.print(readRed()); Serial.print("\t");
  Serial.print(readGreen()); Serial.print("\t");
  Serial.print(readBlue()); Serial.print("\t");
  Serial.println(readWhite());
}
```

To try out the project, place it next to objects of different colors. Open the serial monitor, and the sensor will distinguish between red, green, and blue objects.

Summary

The test program could easily be extended to allow a greater range of colors to be detected.

PART TWO

Security

CHAPTER 5

RFID Door Lock

Difficulty: ★	Cost guide: $60

Radiofrequency identification (RFID) readers are often used in commercial access control as an alternative to keys. They are also sometimes referred to as *near-field communication* (NFC) readers, and they can read data from a tag without the need for any electrical contacts. You just place the tag near the reader, and it will read a unique ID from the tag. The tag may be in the form of a credit card or key fob. Figure 5-1 show a simple homemade RFID lock.

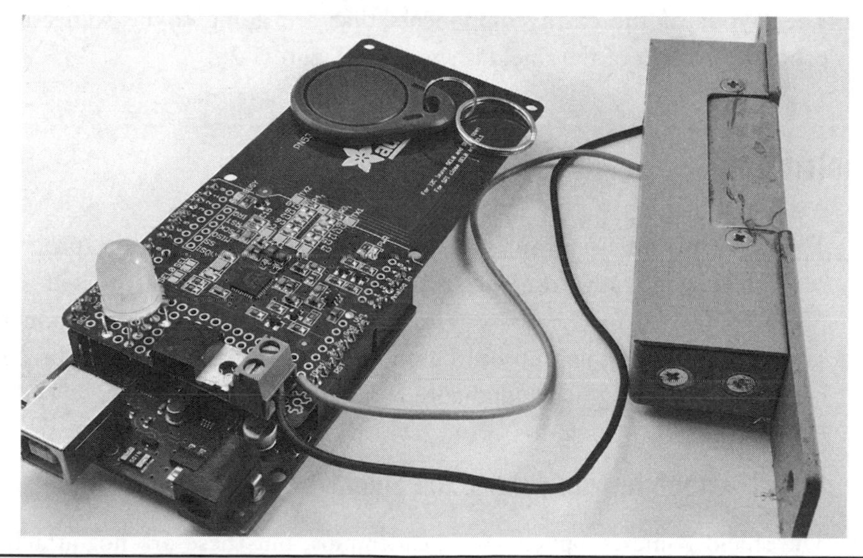

Figure 5-1 A homemade RFID door lock.

Parts

This project uses a ready-made RFID shield from Adafruit. This shield also has a small prototyping area that is just big enough for a red-green-blue (RGB) LED to indicate status and a power MOSFET transistor to switch the 12 V needed by the door latch.

Note that this shield is not compatible with the Arduino Leonardo, so you will need to use an Arduino Uno for this one.

To build this project, you will need the following:

Name	Quantity	Description	Appendix
	1	Adafruit PN532 NFC/RFID shield	M8
	1	12 V door latch	M9
R1–3	3	270 Ω, ¼ W resistors	R1
R4	1	1 kΩ, ¼ W resistor	R2
LED	1	RGB common-cathode LED	S9
T1	1	Power MOSFET	S7
	1	Screw terminal, two way, 0.2 inch	H7
	1	12 V, 2 A power supply	M3

Shield Layout

The layout of the extra components that are going to be soldered onto the prototyping area of the shield is shown in Figure 5-2.

Construction

The construction using an RFID shield is very similar to the process using a protoshield. The only difference is that you need to confine yourself to the small prototyping area on the board. However, we will still attach components in the same way, joining them up on the underside of the board using the component leads and extra lengths of solid-core wire where necessary.

Step 1: Attach Header Pins to the Shield

The shield comes supplied with header pins, but these are not attached to the board. The shield also should include a tag or two for testing (Figure 5-3). If you

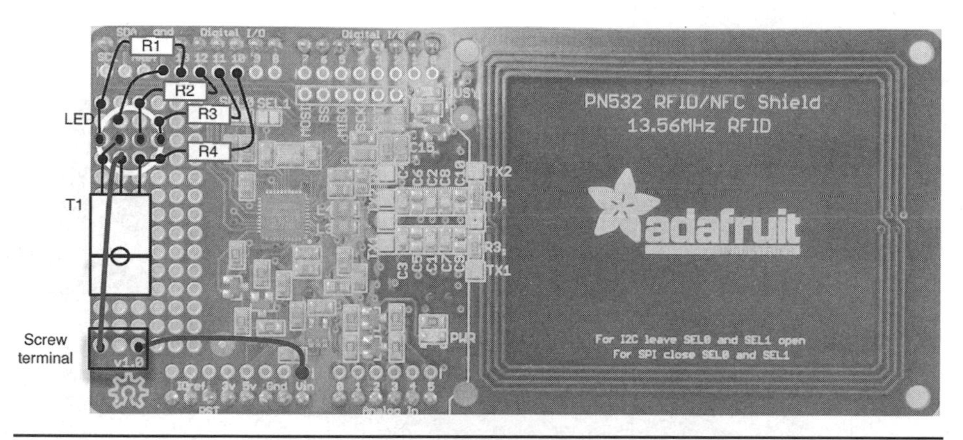

Figure 5-2 Shield layout for the RFID door lock.

want to buy additional cards, make sure that they are of the type 13.56 MHz. There are various different frequencies of tag in use, so be sure to get the right ones.

A good way to make sure that the header pins go in the right holes on the shield and that they are straight is to fit the headers into an Arduino and then place the shield onto the pins for soldering, as shown in Figure 5-4.

Figure 5-3 The Adafruit RFID shield kit.

Figure 5-4 Soldering header pins onto the shield.

Having attached the header leads, you can, if you wish, test the shield to make sure that it is working okay before you attach the rest of the components. The sketch will still work and display the ID of cards that are read if you skip ahead to the section "Software" and then return here to finish the construction.

Step 2: Solder the Resistors

Figure 5-5 shows the resistors soldered into place. In this case, it is easier to push the leads of the resistors through and solder them to the top of the pad.

You can snip off the resistor leads that go to Arduino pins 10 to 13, but leave the other ends of the resistor leads intact for making connections.

Step 3: Solder the Remaining Components

You can now solder on the remaining components, starting with the MOSFET. The leads on this transistor can be carefully bent over so that the MOSFET lies flat on the board (Figure 5-6).

When soldering the LED, make sure that the longest lead (the common cathode) is the second lead in from the left of the board. That is the lead that will be connected to ground and not to a resistor. Also make sure that the screw

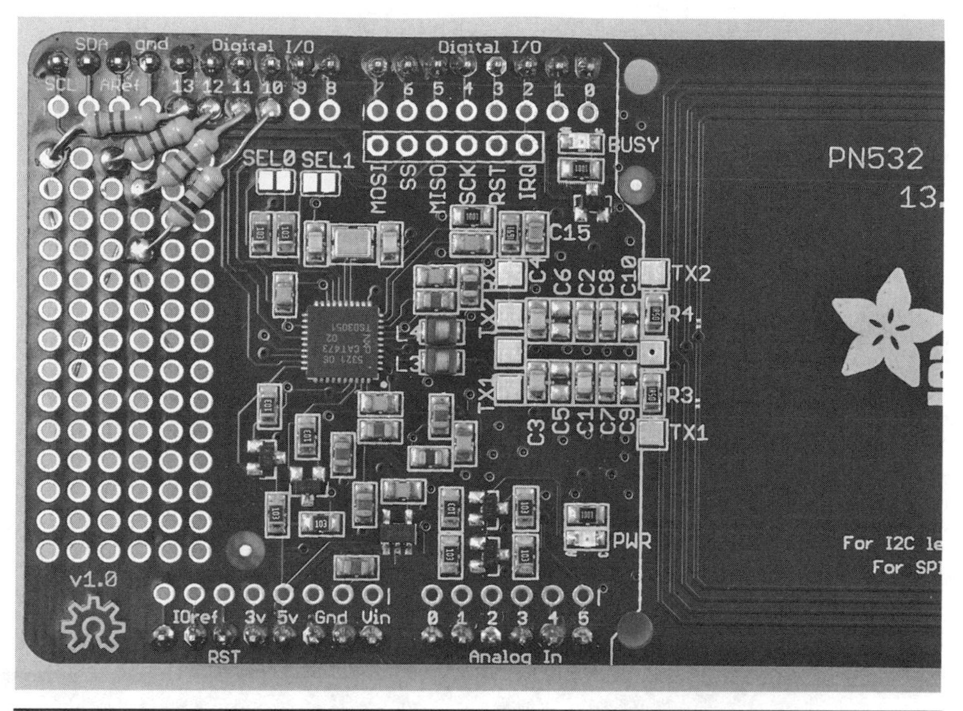

Figure 5-5 Soldering the resistors.

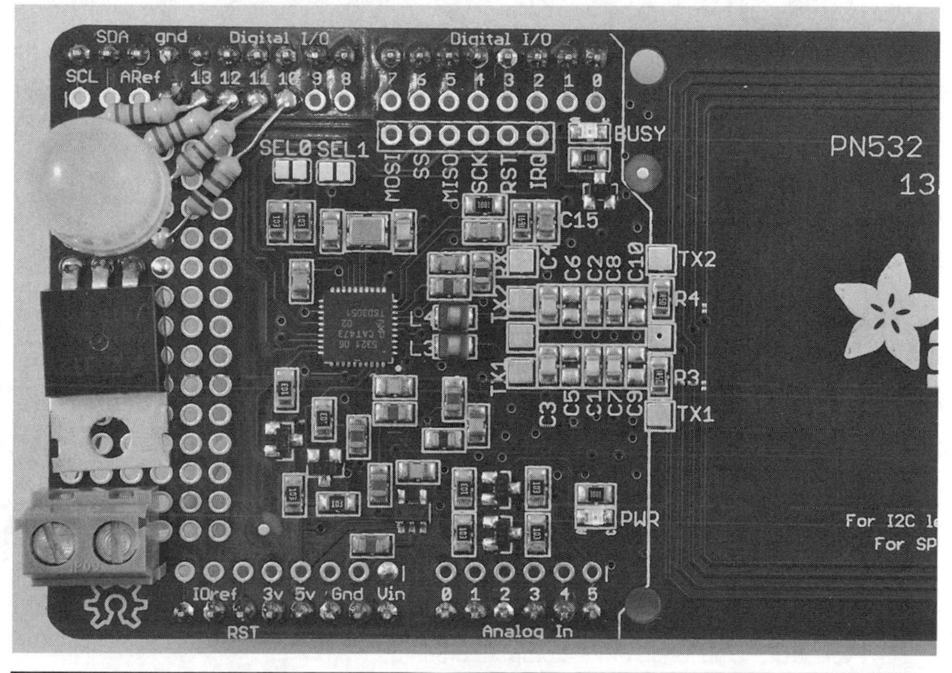

Figure 5-6 Soldering the remaining components.

terminal is attached such that the open end on the side where you insert the wires goes toward the outside of the board.

Step 4: Wire the Underside

Now that all the components are in place, you can make the remaining connections on the underside of the board. Use Figure 5-2 as a reference. When everything is connected, the underside of the shield should look something like Figure 5-7. Note that the board has been flipped top to bottom, so the screw terminal is at the top of the diagram now.

FIGURE 5-7 Underside of the completed board.

Software

The Adafruit RFID shield has an Arduino library written specifically for it. This needs to be downloaded from the following URL: https://github.com/adafruit/Adafruit_NFCShield_I2C. After downloading the zip file, you will need to rename it by removing "master" from the end of the folder name and copying the whole folder into your "Libraries" folder. You will also have to restart the Arduino IDE for it to pick up the new library.

Install the sketch ch_05_rfid onto the Arduino Uno, and attach the shield. Open the serial monitor and set the baud rate to 9,600. When you wave a tag in front of the antenna of the shield, you should see some output such as that shown below in the serial monitor:

```
Scan a tag
Tag Code: 9E2155A7
```

Note that during testing, you can just power the Arduino and shield through the USB connection. Later, when you deploy the lock for real, you will need to use the 12 V power supply to provide enough voltage and current to operate the door latch.

The ID of the card read is shown as a series of hexadecimal digits. If you wish, you could now also load up one of the example sketches that are provided with the Adafruit library. A good choice would be iso14443a_uid. This will display a bit more information about the tag that is scanned and provide more feedback about the shield.

Later, you will use the display in the serial monitor of the tag code to set the "allowed" tags for the project. If you were just running this sketch to check that the reader is working, then you can return to the preceding section and finish the construction. If, on the other hand, you have finished the soldering, then you can now take a look at how the code works.

After including the libraries needed (`Wire` is for I2C communication), there is a constant for the maximum number of codes allowed. If you need more than this, change its value and also the number of entries in the array that follows it. Note that you can leave entries blank if they are not in use.

```
#include <Wire.h>
#include <Adafruit_NFCShield_I2C.h>

const int numCodes = 5;
```

```
// replace with your codes
const char *codes[] = {
  "AD640DA4",
  "FB800DD5",
  "9E2155A7",
  "",
  ""
};
```

Following this are constants for the pins used and also for the colors red, green, and blue.

```
const int redPin = 13;
const int greenPin = 12;
const int bluePin = 11;
const int lockPin = 10;

const int red = 0b001;
const int green = 0b010;
const int blue = 0b100;
```

The following line of code initializes the RFID shield using pin 2 for interrupt requests and pin 3 as the reset pin:

```
Adafruit_NFCShield_I2C nfc(2, 3); // IRQ, RESET pins
```

This is followed by declaration of the global variable `code` that is used to contain the ID of the last tag scanned.

```
char code[16]; // hex string of code
```

The `setup` function initializes the outputs, sets the LED color to blue, starts serial communication, and configures the RFID shield.

```
void setup(void)
{
  pinMode(redPin, OUTPUT);
  pinMode(greenPin, OUTPUT);
  pinMode(bluePin, OUTPUT);
  pinMode(lockPin, OUTPUT);
  setColor(blue);

  Serial.begin(9600);
```

```
  nfc.begin();
  nfc.setPassiveActivationRetries(0xFF);
  nfc.SAMConfig(); // configure board to read RFID tags
  Serial.println("Scan a tag");
}
```

The main `loop` function is very light; it invokes a function `scanCode` that attempts to scan a tag. If this is successful, it returns true and then calls `checkCode`, which will compare the code just scanned with the stored codes.

The function `scanCode` does the actual work of reading the code and converting it into a string of hexadecimal characters.

```
boolean scanCode()
{
  boolean success;
  byte uid[] = { 0, 0, 0, 0, 0, 0, 0 };  // Buffer to store the
                                         // returned UID
  byte uidLength;  // Length of the UID (4 or 7 bytes depending
                   // on ISO14443A card type)

  success = nfc.readPassiveTargetID(PN532_MIFARE_ISO14443A,
    &uid[0], &uidLength);

  if (success)
  {
    Serial.print("Tag Code: ");
    for (uint8_t i=0; i < uidLength; i++)
    {
      char hexDigits[3];
      sprintf(hexDigits, "%02X", uid[i]);
      code[i*2] = hexDigits[0];
      code[i*2+1] = hexDigits[1];
    }
    code[uidLength*2] = '\0';
    Serial.println(code);
  }
  return success;
}
```

The function `scanCode` reads the ID into byte array (`uid`). It then iterates over each byte, converting it into a two-digit hexadecimal string that is then added to the code variable. The function also reports the currently read code in the serial monitor of the Arduino IDE so that you can copy and paste the tag codes from the serial monitor back into the codes array to grant access to new tags.

The checkCode function compares the last scanned tag ID with each of the codes in the codes array. If it finds a match, it sets a flag to that effect. After all the codes have been checked, if a match was found, the function unlockDoor is called; otherwise, the LED is set red for 2 seconds to indicate the attempted use of an invalid tag.

```
void checkCode()
{
  boolean codeValid = false;
  for (int i = 0; i < numCodes; i++)
  {
    if (strcmp(code, codes[i]) == 0)
    {
      codeValid = true;
    }
  }
  if (codeValid)
  {
    unlockDoor();
  }
  else
  {
    setColor(red);
    delay(2000);
    setColor(blue);
  }
}
```

The function unlockDoor sets the LED to be green, activates the lock to allow access, waits 5 seconds, and then locks the door again before setting the LED back to blue.

```
void unlockDoor()
{
  setColor(green);
  digitalWrite(lockPin, HIGH);
  delay(5000); // 5 seconds to get in
  digitalWrite(lockPin, LOW);
```

```
  setColor(blue);
}
```

Finally, the function `setColor` sets the color of the LED from the color supplied as an argument. This is represented as just a three-bit number, with one bit for each of the three channels.

```
void setColor(int color)
{
  digitalWrite(redPin, bitRead(color, 0));
  digitalWrite(greenPin,  bitRead(color, 1));
  digitalWrite(bluePin,  bitRead(color, 2));
}
```

Installing and Using the Door Lock

The type of latch used is designed to replace the receptacle for a conventional lock so that when power is applied, it allows the receptacle to swivel, and the latch can then open. The lock is opened using a solenoid that requires a few hundred milliamps to activate. This is why the Arduino needs to be powered from a 12-V, 2-A power supply. This power is also provided via the shield and MOSFET through the screw terminals, so the two leads connected to the coil of the latch mechanism can just be connected to the screw terminals on the shield.

To set the lock up for your tags, you will need to use it connected to your computer with the Arduino serial monitor open so that you can copy the IDs of your tags into the sketch and then upload the sketch onto the Arduino. One obvious improvement to this lock would be to allow the lock to be "trained" to accept new cards. In this case, the codes would need to be stored in EEPROM memory on the Arduino so as to be preserved should the power be lost to the Arduino.

In normal use, the LED will be blue when the lock is ready. If a known tag is placed near it, the LED will go green, and the lock will open for 5 seconds. If the tag is not known, the LED will go red for 2 seconds.

Summary

In Chapter 6 we will build another lock project that uses a numeric keypad rather than an RFID reader.

CHAPTER 6
Keypad Door Lock

Difficulty: ★★	Cost guide: $40

This project is very similar to the preceding project. However, instead of using a radiofrequency identification (RFID) tag reader, this version of the lock uses a traditional keypad (Figure 6-1).

Parts

This project uses a Protoshield and a red-green-blue (RGB) LED to indicate status and a power MOSFET transistor to switch the 12 V needed by the door latch.

To build this project, you will need the following:

Part	Quantity	Description	Appendix
	1	Protoshield bare printed circuit board (PCB)	A3
	1	Keypad	C2
	1	Stackable header kit[a]	H8
	1	Header strip, 40 way (for keyboard and Arduino)[a]	H2
	1	12 V door latch	M9
R1–3	3	270 Ω, ¼ W resistor	R1
R4	1	1 kΩ, ¼ W resistor	R2
LED	1	RGB common-cathode LED	S9
T1	1	Power MOSFET	S7

(continued on next page)

	1	Piezo buzzer	C3
	1	Screw terminal, two way, 0.2 inch	H7
	1	12 V, 2 A power supply	M3

[a]To keep the wiring as simple as possible, the keypad fits directly into a stackable header pin set on digital pins 0–7. You will probably have to buy a whole set of stackable headers rather than just one.

Protoshield Layout

Figure 6-2 shows the Protoshield layout for this project.

There are a couple of important things to note about this that result from the keypad being connected directly into pins D1 to D7 of the Arduino:

Figure 6-1 Keypad door lock.

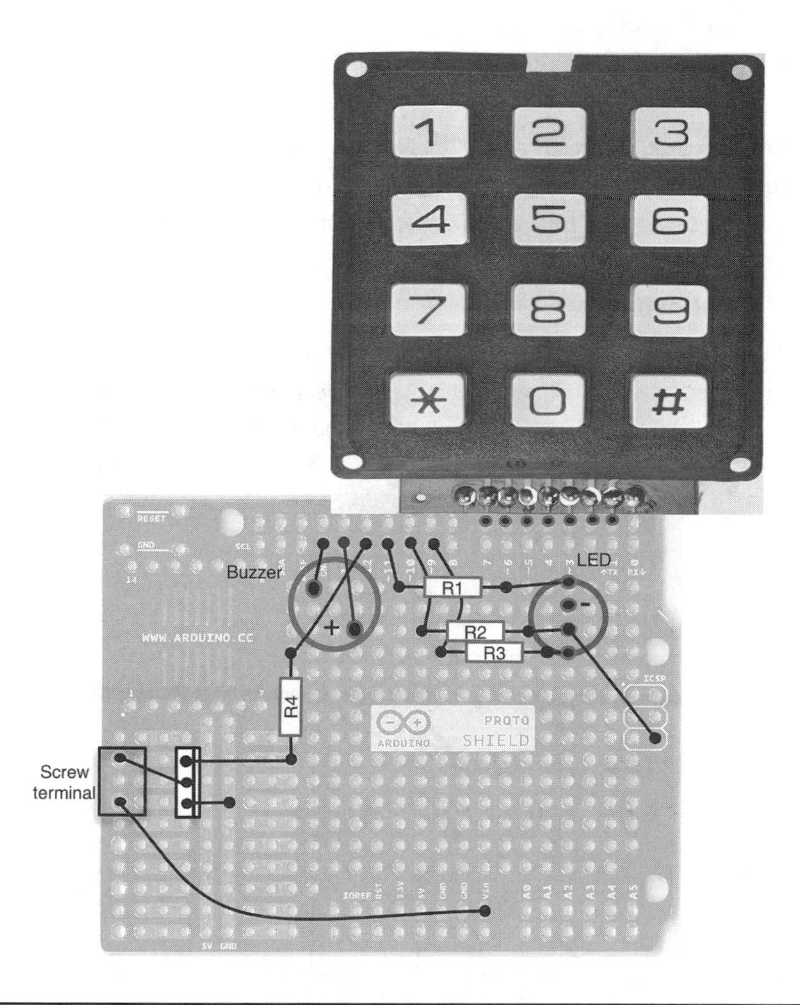

FIGURE 6-2 Protoshield layout for the keypad door lock.

1. The keypad has one extra pin on either side of the seven middle pins that are actually used.
2. This means that one of the pins of the keyboard is connected to Arduino pin D1 (Tx). This is normally a pin that we avoid because on an Arduino Uno it is needed by the USB port to program the Arduino. Thus, if you are using a Uno, you will have to unplug the keypad while programming the board. If, on the other hand, you are using a Leonardo, then you can leave it connected because the Leonardo uses a separate interface for the USB.

Construction

This project is pretty easy to assemble, but there are a few cases where bare component leads will need to hop over other tracks on the prototshield without touching them, so follow the construction steps with care.

Matrix Keypads

Keypads are normally arranged in a grid so that when one of the keys is pressed, it connects a row to a column. Figure 6-3 shows a typical arrangement for a 12-key keyboard with numbers from 0 to 9 and * and # keys.

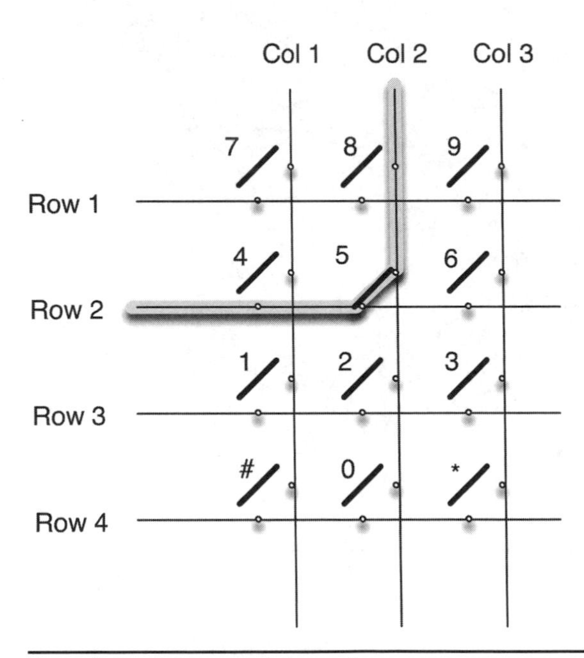

FIGURE 6-3 A 12-switch keypad.

The key switches are arranged at the intersection of row-and-column wires. When a key is pressed, it connects a particular row to a particular column. By arranging the keys in a grid such as this, we only need to use seven (four rows + three columns) of our digital pins rather than 12 (one for each key).

However, this also means that we have to do a bit more work in the software to determine which keys are pressed. The basic approach we have to take is to connect each row to a digital output and each column to a digital input. We then put each output `HIGH` in turn and see which inputs are `HIGH`.

Step 1: Attach Header Pins to Keypad

The keypad uses seven pins but has a spare connector at each end that is not connected to anything. You only need to solder pins to the seven pins, but you can put the extra pins on each end if you prefer. Figure 6-4 shows my keypad. Note that I have chosen to include the extra unused pins at each end of the keypad connector.

Step 2: Attach Header Pins to the Shield

In this project, we will not be using standard header pins for every pin on the Protoshield, but rather we will be using a stackable header for pins 0–7 so that the keyboard can just be plugged directly into it. Start by soldering this stackable header onto pins 0–7. Push the header through from the top and solder it on the underside of the shield, as shown in Figure 6-5. The remainder of the headers can be normal header pins. Figure 6-6 shows these soldered in place.

FIGURE 6-4 Soldering pins onto the keypad.

FIGURE 6-5 Soldering the stackable header.

FIGURE 6-6 Soldering the normal headers.

Step 3: Solder the Resistors

Figure 6-7 shows the resistors soldered into place. In this case, it is easier to push the leads of the resistors through and solder them to the top of the pad.

You can snip off the resistor leads that go to the Arduino pins but leave the other ends of the resistor leads intact for making connections.

Step 4: Solder the Remaining Components

You can now solder the remaining components, starting with the MOSFET. The leads on this transistor can be carefully bent over so that it lies at an angle to the board (Figure 6-8). Make sure that the metal parts of the transistor package are not touching any of the solder pads on the Protoshield.

When soldering the LED, make sure that the longest lead (the common cathode) is the second lead down from the top of the board. This is the lead that will be connected to ground and not to a resistor. Also make sure that the screw terminal is attached with the open end, where the wires go toward the outside of the board.

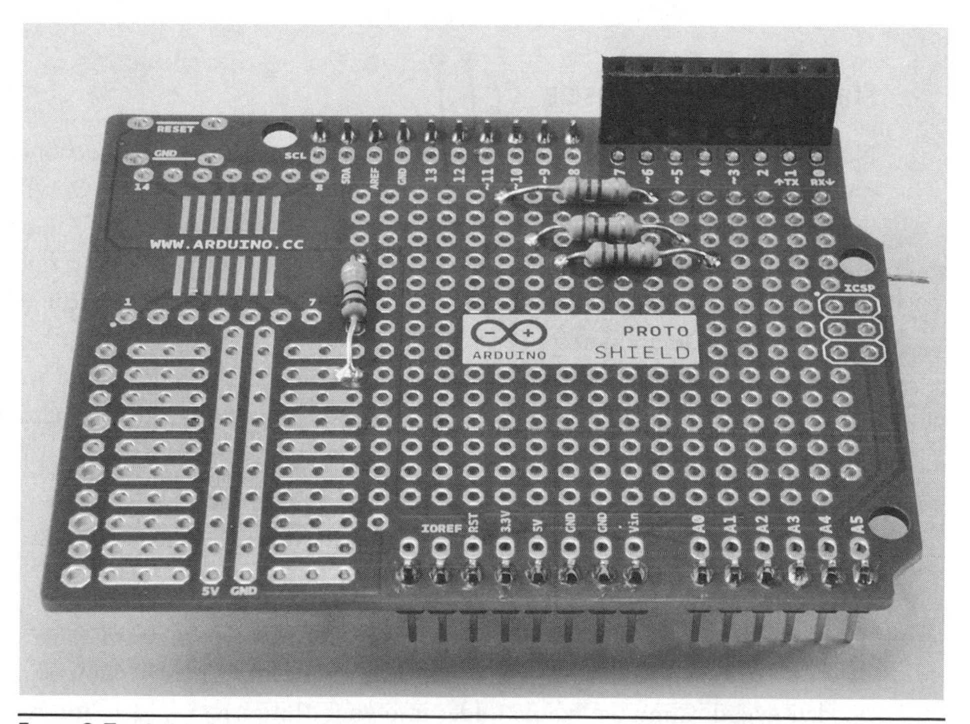

FIGURE 6-7 Soldering the resistors.

Figure 6-8 Soldering the remaining components.

Step 5: Wire the Underside

Now that all the components are in place, you can make the remaining connections on the underside of the board. Use Figure 6-2 as a reference. When everything is connected, the underside of the shield should look something like Figure 6-9. Figure 6-10 shows a diagrammatic version of the underside for easy reference. Note that the board has been flipped left to right so that the screw terminal is at the right of the diagram now.

You will notice that near the screw terminal and the MOSFET, there are some links made with the component leads that need to jump over other solder pads without making contact.

Software

There is an Arduino library that greatly simplifies using a keypad. This library needs to be downloaded from the following URL: http://playground.arduino.cc/Code/Keypad. Copy the whole folder into your "Libraries" folder. You will also have to restart the Arduino IDE for it to pick up the new library.

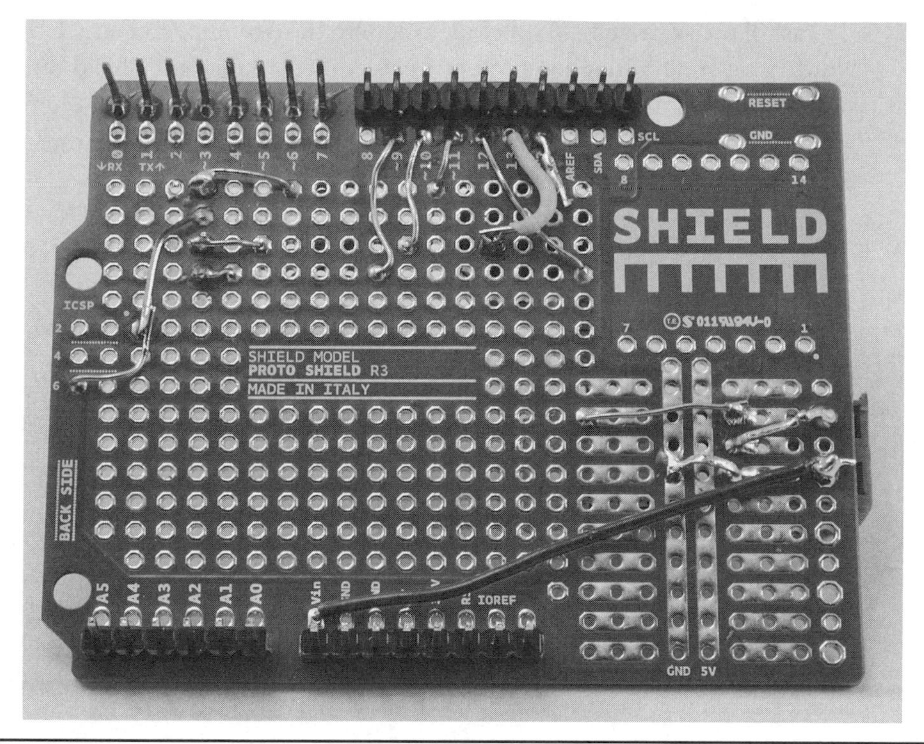

FIGURE 6-9 Underside of the completed board.

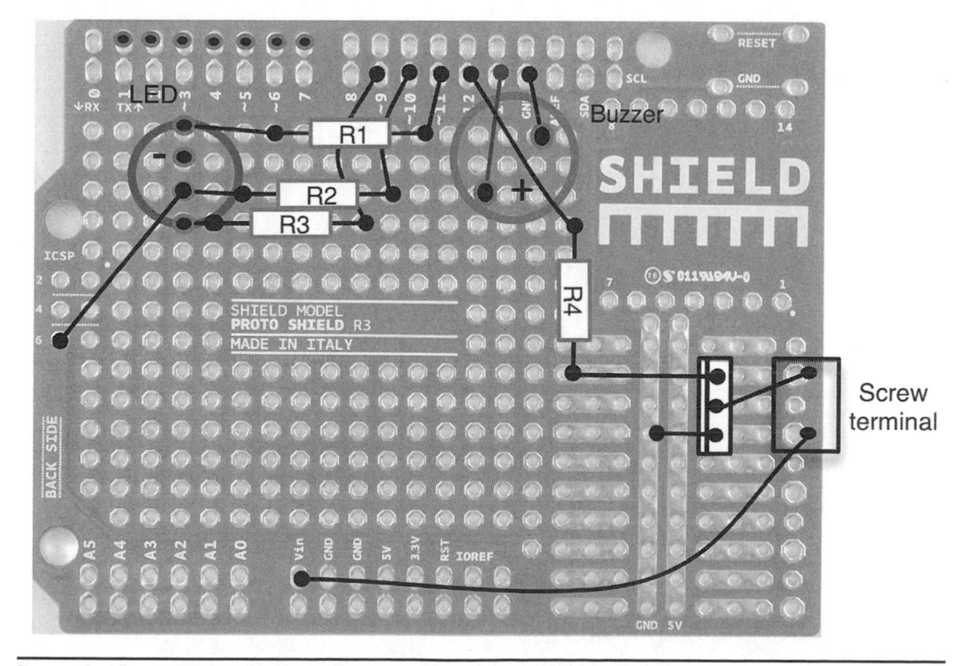

FIGURE 6-10 Wiring diagram for the underside of the board.

Install the sketch ch_06_keypad_lock onto the Arduino, and attach the shield. Note that during testing, you can just power the Arduino and shield through the USB connection. Later, when you deploy the lock for real, you will need to use the 12-V power supply to provide enough voltage and current to operate the door latch.

Much of the code for this project is very similar to that of the RFID door lock of Chapter 5, so you also may wish to refer to that chapter. Here I will just present the parts of the sketch that are different.

The first difference is that we need to include the "Keypad" library at the top of the file.

```
#include <Keypad.h>
```

We then have a variable that keeps the secret code. To change this lock's code, you will need to change this and then reupload the sketch.

```
char* secretCode = "1234";
```

The following lines set up the "Keypad" library, telling it both what the key labels are for the keypad and also the Arduino pins that will be connected to each row and column of the keypad:

```
const byte rows = 4;
const byte cols = 3;
char keys[rows][cols] = {
  {'1','2','3'},
  {'4','5','6'},
  {'7','8','9'},
  {'*','0','#'}
};
byte rowPins[rows] = {6, 1, 2, 4};
byte colPins[cols] = {5, 7, 3};
Keypad keypad = Keypad(makeKeymap(keys), rowPins, colPins, rows,
  cols);
```

This sketch uses the same constants and functions for controlling the color of the RGB LED, so I will skip over these.

The `loop` function is as follows:

```
void loop()
{
  boolean wrong = false;
  while (waitForKey() != '*') {};
```

```
  for (int i = 0; i < 4; i++)
  {
    setColor(blue);
    char key = waitForKey();
    if (key == '*')
    {
      setColor(red);
      break;
    }
    if (key != secretCode[i]) wrong = true;
  }
  if (!wrong)
  {
    unlockDoor();
  }
  else
  {
    lockDoor();
  }
}
```

Here we first define a Boolean flag called wrong that will be set to true if a wrong digit is entered as the code. We then have a `while` loop that waits for the "*" key to be pressed. Pressing the "*" key is the way that the user indicates that he or she wants to enter the secret code.

Once this key has been pressed, there is a loop that captures the next four key presses and compares them with the appropriate digits of the secret code. If any of the digits are wrong, then the `wrong` flag is set. After all four key presses have been captured, then either the door is unlocked or locked depending on the result in the `wrong` flag.

Unlocking the door is similar to the sketch of Chapter 5, except that in addition to setting the color of the LED, the program also sounds the buzzer continuously for the 5 seconds that the latch remains unlocked.

```
void unlockDoor()
{
  setColor(green);
  tone(buzzerPin, 500);
  digitalWrite(lockPin, HIGH);
  delay(5000); // 5 seconds to get in
```

```
  digitalWrite(lockPin, LOW);
  noTone(buzzerPin);
  lockDoor();
}
```

Sounding the buzzer requires use of the Arduino `tone` command. The first parameter of this command is the pin on which the tone is to be played, and the second is the frequency of the note in hertz. The `noTone` command cancels the buzzing sound at the end of the 5 seconds, after which the door is then locked again using the following `lockDoor` function:

```
void lockDoor()
{
  setColor(red);
  digitalWrite(lockPin, LOW);
  for (int i = 0; i < 5; i++)
  {
    setColor(0);
    tone(buzzerPin, 1000);
    delay(500);
    setColor(red);
    noTone(buzzerPin);
    delay(500);
  }
}
```

The `lockDoor` function also gets called as a result of entering an incorrect code. Therefore, to make the failure to enter a correct code obvious, a `for` loop is used to make the LED flash red and the buzzer sound five times.

The final function that I need to mention is `waitForKey`.

```
char waitForKey()
{
  char key;
  while ((key = keypad.getKey()) == 0) {}; // wait for key down
  while (! keypad.getKey() == 0) {}; // wait for key up
  delay(10);
  return key;
}
```

This function will wait until a key has been both pressed and released. The first `while` loop waits until a key is pressed. The second `while` loop waits until the key is released again. The `delay` helps to prevent false key presses as a result of key bounce.

Installing and Using the Door Lock

The type of latch used is designed to replace the receptacle for a conventional lock so that when power is applied, it allows the receptacle to swivel, and the latch can then open. The lock is unlocked using a solenoid that requires a few hundred milliamps to activate. This is why the Arduino needs to be powered from a 12 V 2 A power supply. This power is also provided via the shield and MOSFET through the screw terminals, so the two leads connected to the coil of the latch mechanism can just be connected to the screw terminals on the shield.

The LED will be red when the lock first starts up. Note that the buzzer will warble for a few seconds if you are using an Arduino Leonardo. Once reset is complete, `setup` calls the `lockDoor` function, so the LED will flash red and the buzzer will sound four times.

To unlock the door, press the "*" key, and the LED will now turn blue. Enter the four-digit code (1234), and the LED will go green, the buzzer will sound, and the door latch will be released for 5 seconds, after which the lock sequence will run again.

Summary

In Chapter 7 we will build the final lock in this book—this time using a buzzer as a "knock" sensor so that we can unlock the door with a secret knock.

CHAPTER 7

Secret Knock Lock

Difficulty: ★★	Cost guide: $35

This project is also similar to the two preceding projects. However, instead of using a radiofrequency identification (RFID) tag reader or keypad, this version of the lock uses a secret knock (Figure 7-1).

The idea is that you train the lock with a certain pattern of knocks, and then, when it hears the correct knock, it releases the door. This type of lock should not be considered to be as secure as the preceding two lock projects but is a great bit of fun.

Parts

Just as with the preceding two projects, this project uses a Protoshield and an RGB LED to indicate status and a power MOSFET transistor to switch the 12 V needed by the door latch.

To build this project, you will need the following:

Part	Quantity	Description	Appendix
	1	Protoshield bare printed circuit board (PCB) and header pins	A3
	1	Header strip, 34 way (Arduino and connection to buzzer 2)	H1
	2	Female-to-female header leads[a]	H9
	1	12 V door latch	M9
R1–3	3	270 Ω, ¼ W resistor	R1
R4	1	1 kΩ, ¼ W resistor	R2
R5	1	4.7 MΩ, ¼ W resistor	R4
LED	1	RGB common-cathode LED	S9
T1	1	Power MOSFET	S7
	2	Piezo buzzer[b]	C3
	1	Screw terminal, two way, 0.2 inch	H7
	1	12 V, 2 A power supply	M3

[a] *Only two female-to-female header leads are needed, but these are normally sold as a set of 10 or more. Alternatively, the second buzzer could be soldered directly to the Protoshield, but this could make it difficult to position it where it can detect the knocks.*

[b] *Note that this project requires two piezo buzzers, one as a buzzer and one to sense knocks.*

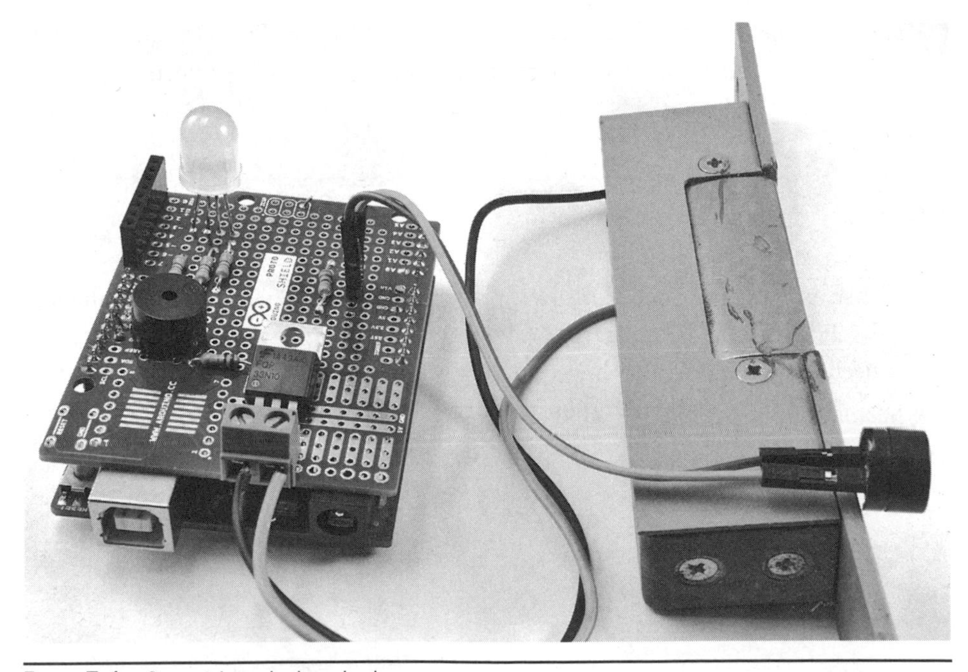

Figure 7-1 Secret knock door lock.

Protoshield Layout

Figure 7-2 shows the Protoshield layout for this project.

Most of this design is exactly the same as that of the keypad lock in Chapter 6. The only differences are

- You do not need to use stackable headers at all. All the headers can be normal 0.1-inch header pins.
- There is a new component in R5 and a pair of header pins to which the second (knock-detecting) buzzer is attached using female-to-female jumper leads.

Construction

Start by following the instructions for constructing the Protoshield in Chapter 6, but use normal header pins throughout rather than stackable headers. If you already made the Chapter 6 project, then simply unplug the keypad and add the extra component (R5) as described in the following steps. Note that even after you have added the extra parts, the hardware will still be compatible with the keypad lock project of Chapter 6.

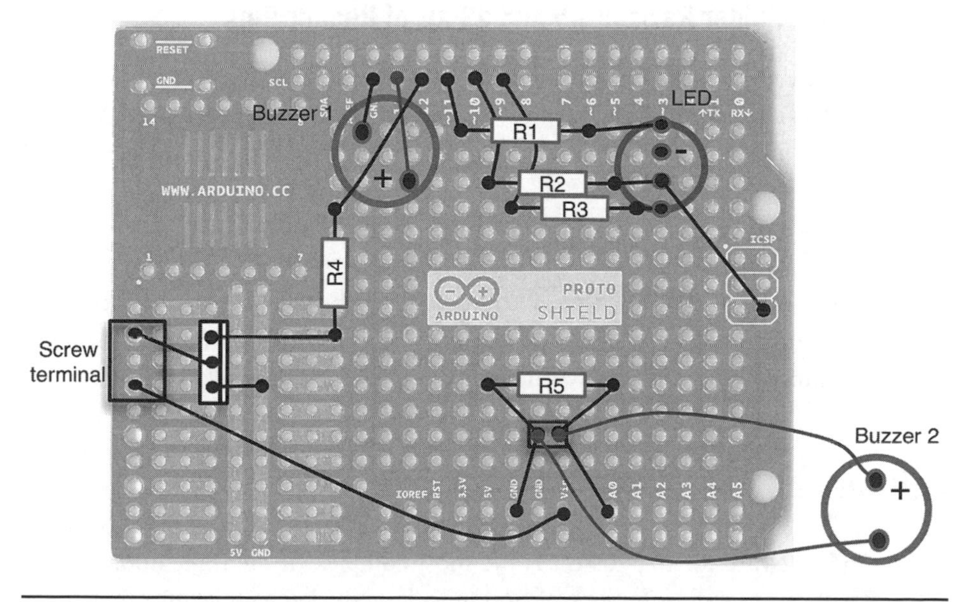

FIGURE 7-2 Protoshield layout for the secret knock door lock.

Figure 7-3 Attaching the new resistor and header pins.

Step 1: Solder Resistor R5 and a Pair of Header Pins

Solder the resistor in first, leaving its leads intact, because they will reach down to the GND and A0 Arduino pins on the Protoshield. Then solder in the pair of header pins. Figure 7-3 shows the additional components on the Protoshield.

Step 2: Wire the Underside

Now bend the resistor leads so that they brush the edge of each header pin and continue on to the Arduino GND and A0 connections on the Protoshield. Figure 7-4 shows the additional wiring on the underside of the Protoshield.

Software

The software for this project is the most complex of the three door lock projects. The basic functions for locking and unlocking the latch and controlling the color of

Figure 7-4 Underside of the completed board.

the LED are the same, but detecting the knock pattern requires the timings of each tap on the sensor (second buzzer) to be stored in the two arrays `key` and `guess`.

```
const int maxTaps = 30;
long key[maxTaps];
long guess[maxTaps];
```

The section of constants above this might need changing to suit your particular buzzer. See the next section if you need to do this.

When the Arduino restarts, the setup function will turn the LED blue, expecting you to immediately enter the secret knock key. The function records this key using the `recordKnock` function and then plays it back using the `playKnock` function.

```
void setup()
{
  pinMode(redPin, OUTPUT);
  pinMode(greenPin, OUTPUT);
  pinMode(bluePin, OUTPUT);
  pinMode(lockPin, OUTPUT);
  setColor(blue);
  recordKnock(key);
  playKnock(key);
  delay(500);
  lockDoor();
}
```

The `recordKnock` function is as follows:

```
void recordKnock(long buffer[])
{
  //Serial.println("Recording for 5 seconds");
  int i = 1;
  long t0 = millis();
  long t = 0;
  while (((t = millis()) < t0 + maxRecordTime) && (i < maxTaps))
  {
    if (tapDetected())
    {
      Serial.print(".");
      buffer[i] = t - t0;
      i++;
    }
  }
  buffer[0] = i;
}
```

This function will write the knock timings into whatever array is passed to it as an argument. This allows the function to be used both for recording the secret key and for capturing the "guess" that you tap in. It uses the first element of the array to record the number of taps detected, and the remainder of the elements of the array will contain the duration in milliseconds after the start of recording (`t0`).

The while loop ensures that the `maxRecordingTime` and `maxTaps` are not exceeded because this would cause unpredictable results if the array bounds were exceeded and fixes the recording time at 5 seconds. The function `recordKnock` relies on the function `tapDetected` to determine whether the sensor has been tapped.

```
boolean tapDetected()
{
  if (analogRead(knockPin) > threshold)
  {
    delay(ignorePeriod); // ignore ringing
    return true;
  }
  return false;
}
```

The function `tapDetected` returns true if the reading on the analog input A0 is above the threshold. However, it delays for `ignorePeriod` to ensure that the ringing effect that might cause a knock to be detected twice is minimized.

Playing back the knock is a similar process to recording it. The function `playKnock` is as follows:

```
void playKnock(long buffer[])
{
  long t0 = millis();
  long prevTap = 0;
  for (int i = 1; i < buffer[0]; i++)
  {
    delay(buffer[i] - prevTap);
    tone(buzzerPin, 500);
    delay(10);
    noTone(buzzerPin);
    prevTap = buffer[i];
  }
}
```

To play back the knocks through the buzzer, we need to step over each element in the array, make a short buzz, and create a delay for the time difference between the current knock and the previous one.

The main `loop` function is as follows:

```
void loop()
{
  if (tapDetected())
  {
    delay(100);
    setColor(blue);
    delay(100);
    recordKnock(guess);
    if (guessCorrect())
    {
      unlockDoor();
    }
    else
    {
      lockDoor();
    }
```

```
    }
}
```

Nothing will happen in the main loop until a tap is detected. This then triggers the sequence of events necessary to capture a series of taps on the sensor. The LED is set blue, and then the `recordKnocks` function is called. The `if` statement uses the `guessCorrect` function to check whether the knocked sequence is close enough to the saved pattern. If it is, then the door is unlocked.

The `guessCorrect` function is as follows:

```
boolean guessCorrect()
{
  if (key[0] != guess[0])
  {
    //Serial.println("wrong number of knocks");
    return false;
  }
  // find delay before first tap
  long startGap = guess[1] - key[1];
  for (int i = 1; i < key[0]; i++)
  {
    long error = abs(guess[i] - key[i] - startGap);
    if (error > tapLeeway)
    {
      return false;
    }
  }
  return true;
}
```

The first thing that happens in this function is that the lengths of the `guess` and `key` knocks are compared. If they are not the same length, then they cannot match, so `false` is returned. If they are of the same length, then the time of each tap has to be compared in turn. Because the knocking is unlikely to start at the same time after the LED went blue, the `startGap` variable is set to a valuc to synchronize the two sequences of knocks. The error of a particular guess is calculated as the absolute value of the difference between the `guess` knock and the `key` knock less the `startGap`. If that error is greater than the limit (in milliseconds) specified in `tapLeeway`, then `false` is returned.

Assuming that all the timings are a close enough match, then the code will fall off the end of the loop, and `true`" will be returned to indicate a match.

Installing and Using the Door Lock

The type of latch used is designed to replace the receptacle for a conventional lock so that when power is applied, it allows the receptacle to swivel and the latch to open. The lock is opened using a solenoid that requires a few hundred milliamps to activate. This is why the Arduino needs to be powered from a 12 V, 2 A power supply. This power is also provided via the shield and MOSFET through the screw terminals, so the two leads connected to the coil of the latch mechanism can just be connected to the screw terminals on the shield.

There are a number of constants at the top of the sketch that can be tweaked to alter the performance of the lock.

```
const int threshold = 3;
const int ignorePeriod = 60; // milliseconds
const int tapLeeway = 200; // milliseconds
const int maxRecordTime = 5000;
```

The constant `threshold` is the value of analog reading above which the reading will count as a knock. If you find that the lock is too sensitive and registers knocks when there were none, then increase this value. The `ignorePeriod` value will help to eliminate double taps where there was only one. Increase this value if you find this happening.

When comparing the knocks, the value of `tapLeeway` specifies the number of milliseconds that the guessing taps can deviate in timing from the `key` taps. If you want to make the lock more lax, then increase this value.

Summary

In Chapter 8 we will look at using a sound shield to play the sound of a dog barking when a panic button is pressed or motion is detected using a passive infrared sensor.

CHAPTER 8
Fake Dog

Difficulty: ★	Cost guide: $35

This project plays a sound file whenever a button is pressed or a passive infrared (PIR) movement sensor is triggered (Figure 8-1). Unlike most of the projects we have built so far, this project uses a ready-made shield for playing the sound file.

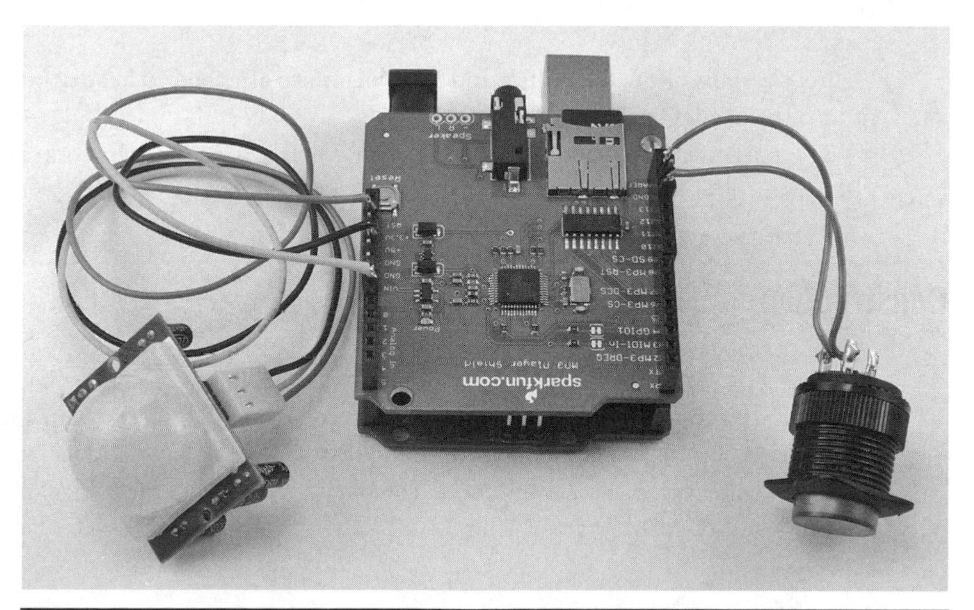

FIGURE 8-1 Fake dog project.

Parts

To build this project, you will need the following:

Part	Quantity	Description	Appendix
	1	Sparkfun MP3 player shield	M25
	1	Set of Arduino shield through headers	H8
	1	PIR sensor module	M5
	1	Large push switch	C1
	1	Powered PC speakers	
	1	Micro Secure Digital (SD) card and SD adapter	
	1	Computer with SD card reader	
	1	Header pin strip, three way + six way	H1

The MP3 player shield is equipped with a 3.5-mm headphone socket. This is ideal for connecting to a pair of powered computer or MP3 player speakers. The louder the better!

Unless you plan to record a very large amount of dog barking, any micro SD card bigger than 128 MB will be fine.

Design

Figure 8-2 shows how the switch and PIR sensor are attached to header pins that fit into the header sockets of the shield.

The push switch grounds the digital input when it is pressed. In contrast, the PIR sensor requires a 5 V and GND supply.

Construction

Because most of the work is already done for us in the ready-made shield, this project is easy to make. The most difficult part probably will be getting a recording of your dog barking. Alternatively, you can download a dog bark sample from the book's website (www.simonmonk.org). The file is contained in the folder "ch_08_barking" and is called "track001.mp3."

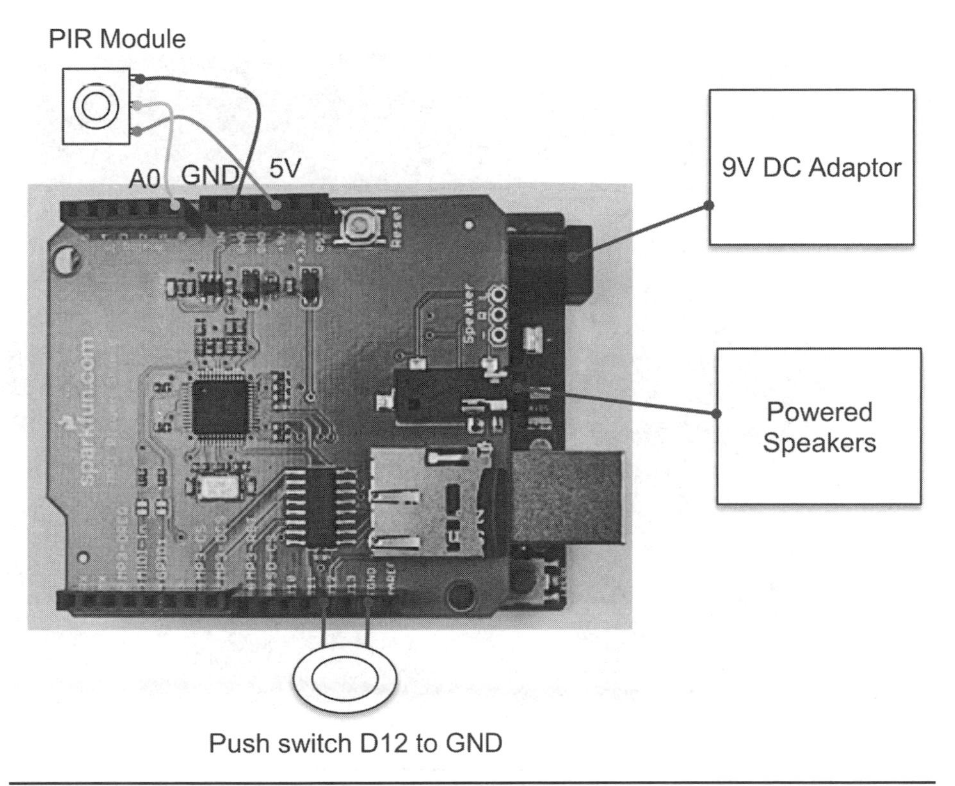

FIGURE 8-2 Fake dog project wiring diagram.

Step 1: Solder the Header Pins to the Shield

The MP3 player shield (Figure 8-3) comes without any header pins attached. Because there are going to be a few extra things that we need to attach to the Arduino, we will use *through header pins* (Figure 8-4).

To solder these in place, start by soldering just one pin on the underside of the board. Then straighten the header row up by melting the solder again while you get it level and flat to the board. Then you can solder the rest of the pins on that header and then repeat the process for the other headers. When all the headers are attached, the finished board should look like Figure 8-5.

Step 2: Attach the Push Button

Solder short leads of multicore insulated wire onto the push button. Then solder the other end of one of the leads to the GND pin indicated in Figure 8-2. Attach the other lead to D12 on the header.

FIGURE 8-3 Sparkfun MP3 player shield.

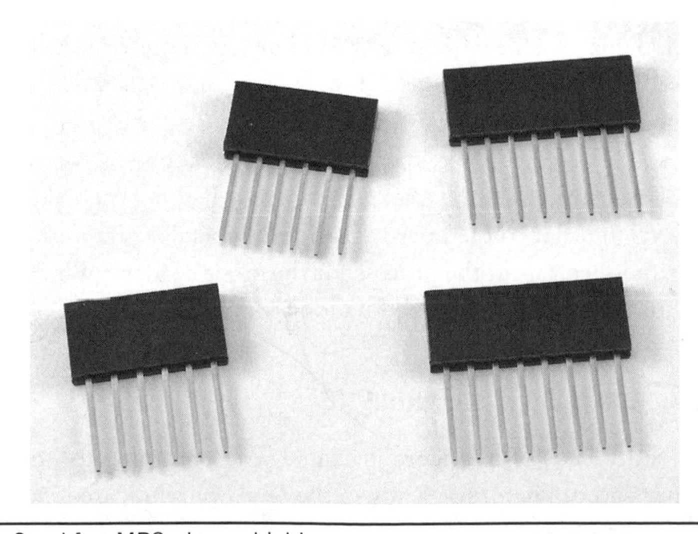

FIGURE 8-4 Sparkfun MP3 player shield.

FIGURE 8-5 Finished shield.

Step 3: Attach the PIR Sensor

The PIR module comes with leads attached. These should be soldered to 5 V, GND, and A0 on the header strip. Again, use Figure 8-2 as a reference.

Step 4: Install the MP3 Player Library

The MP3 shield is a little complex to drive. Data have to be set to it a byte at a time. Fortunately, to simplify this process, a library has been developed that you need to download and install into your Arduino environment.

First, download the the zip file from www.billporter.info/2012/01/28/sparkfun-mp3-shield-arduino-library/. Unzip the file, and you will find that you have a folder called something like "Sparkfun-MP3-Player-Shield-Arduino-Library-master." In this folder, there are two folders, "SdFat" and "SFEMP3Shield," that need to be copied to your "Arduino" folder.

More specifically, they need to be copied into a folder called "Libraries" in your "Arduino" folder within your "Documents" folder. You may have to create a folder called "Libraries" within the "Arduino" folder where all your saved sketches live.

For the libraries to be recognized, you need to restart the Arduino software.

Step 5: Prepare a Micro SD Card

Format the micro SD card in FAT32 format, and then download the file track001 .mp3 from the book's website (www.simonmonk.org) and copy it onto the SD card. You may prefer to record your own deterrent sound. If you do, then this must be in MP3 format recorded at 192 kbs. Remember to name the file track001.mp3.

When the file is ready, remove the micro SD card and insert it into the holder on the MP3 shield.

Using the Fake Dog

Once built and programmed, this project no longer needs to be connected to your computer for anything other than power. Therefore, you probably will want to power it from an external power adapter. The same applies to the speakers.

Ideally, the PIR sensor will be positioned outside your door, with wires running into the house and connecting to the shield.

Software

The library makes the sketch itself nice and simple. The sketch is called ch_08_ barking.

```
#include <SPI.h>
#include <SdFat.h>
#include <SdFatUtil.h>
#include <SFEMP3Shield.h>

int buttonPin = 12;
int pirPin = A0;

SFEMP3Shield MP3player;

void setup()
{
  MP3player.begin();
  MP3player.SetVolume(0, 0);
  pinMode(buttonPin, INPUT_PULLUP);
  pinMode(pirPin, INPUT);
```

```
  delay(5000);
}

void loop()
{
  if (digitalRead(buttonPin) == LOW || digitalRead(pirPin) ==
    HIGH)
  {
    MP3player.playTrack(1);
    delay(1000);
  }
}
```

The key part is the line

```
MP3player.playTrack(1);
```

This tells the library to look for a file called track001.mp3.

The remainder of the code is mostly concerned with checking the digital inputs (one for the PIR and one for the switch).

Summary

This is a fun project to make and could be adapted to other uses. To improve the project, you could record a number of different tracks and have them played at random when triggered.

CHAPTER 9

Person Counter

Difficulty: ★★	Cost guide: $30

This project uses an ultrasonic range finder and an LED display module to count people moving past the range finder and display the number (Figure 9-1).

Parts

This project uses a Protoshield and an RGB LED to indicate status and a power MOSFET transistor to switch the 12 V needed by the door latch.

To build this project, you will need the following:

Part	Quantity	Description	Appendix
	1	Protoshield bare printed circuit board (PCB)	A3
	1	HC-SR-04 ultrasonic range finder	M11
	1	Four-digit, seven-segment LED display[a]	M6 or M10
	1	Header strip, male (for shield)	H1
	1	Header strip, female, four way and five way	H2
S1	1	Tactile push switch	C2

[a]*This project is designed to use Adafruit seven-segment LED displays. It is compatible with both the 0.56-inch display and the superlarge 1.2-inch display. The 1.2-inch display has one additional pin that requires a 5-V supply. Both displays will work just fine in this design.*

Figure 9-1 Person counter.

Protoshield Layout

Figure 9-2 shows the Protoshield layout for this project.

Construction

The project uses header sockets to connect the modules. If you prefer, you can just solder them directly to the Protoshield.

Step 1. Attach Header Pins to the Protoshield

Attach header pins to the Protoshield as normal. However, you may choose to omit the pins for the Arduino pins A0 to A5 as these are not used in this design.

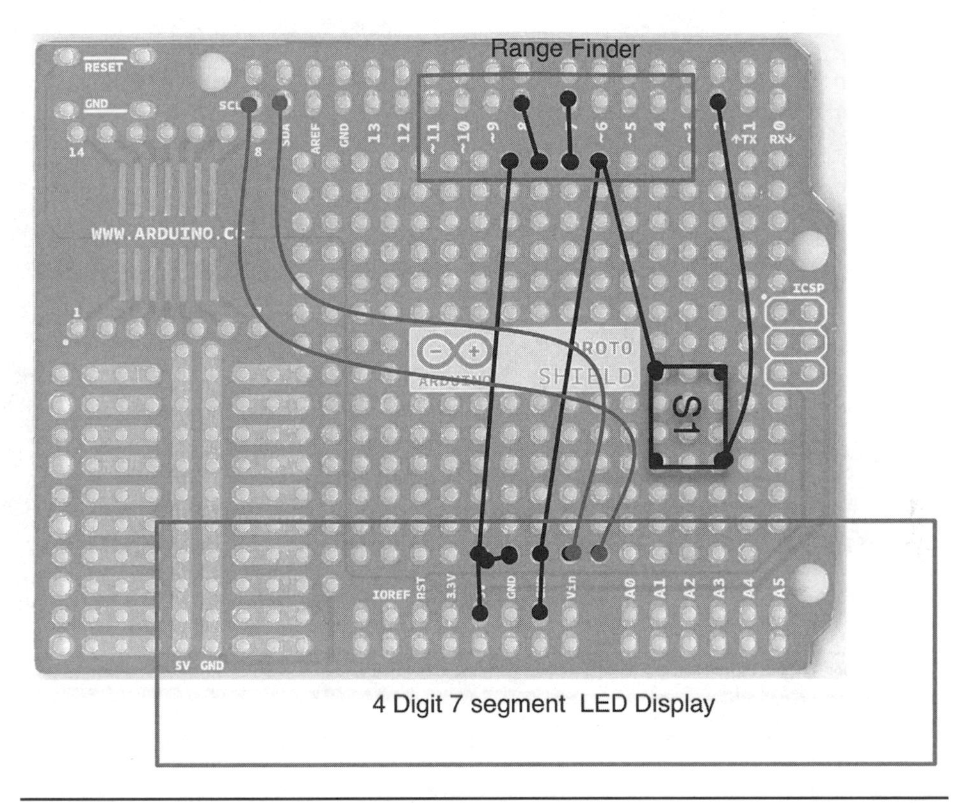

FIGURE 9-2 Protoshield layout for the person counter.

Step 2. Solder the Components to the Protoshield

The header sockets can be cut from a long length of header sockets (see Chapter 3 for details on how to do this). There are only two header sockets and the push switch to attach, so they can all be soldered in one go. Use Figure 9-2 as a reference for positioning the header sockets and the switch. Figure 9-3 shows the top side of the completed Protoshield.

Step 3: Solder the Underside of the Protoshield

Figure 9-4 shows the underside of the board, and Figure 9-5 shows the wiring diagram from the underside of the board. There are a few places where wires cross over each other, so use insulated wire where necessary.

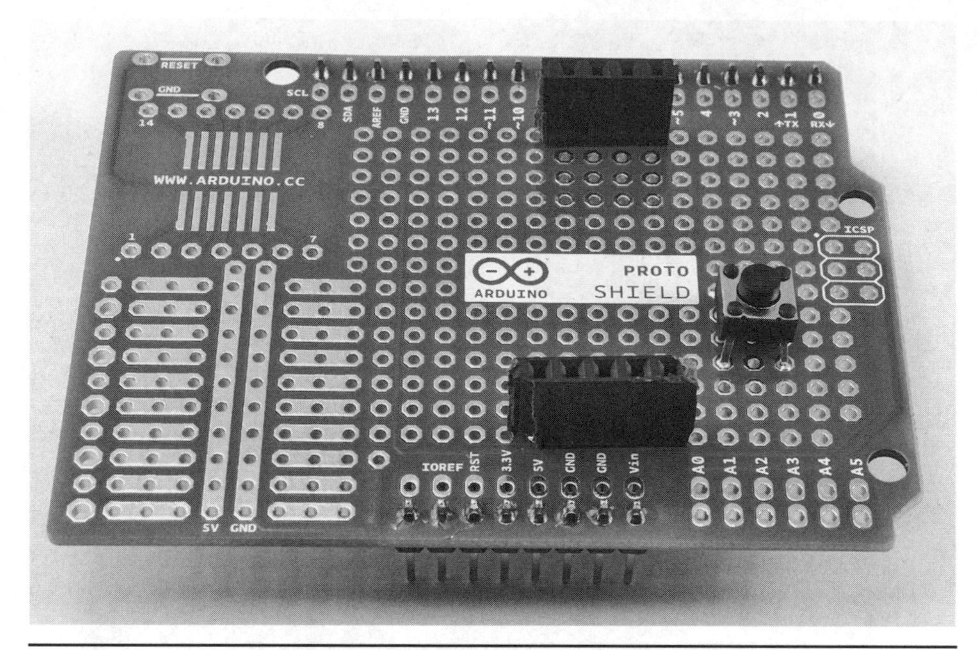

FIGURE 9-3 Completed Protoshield.

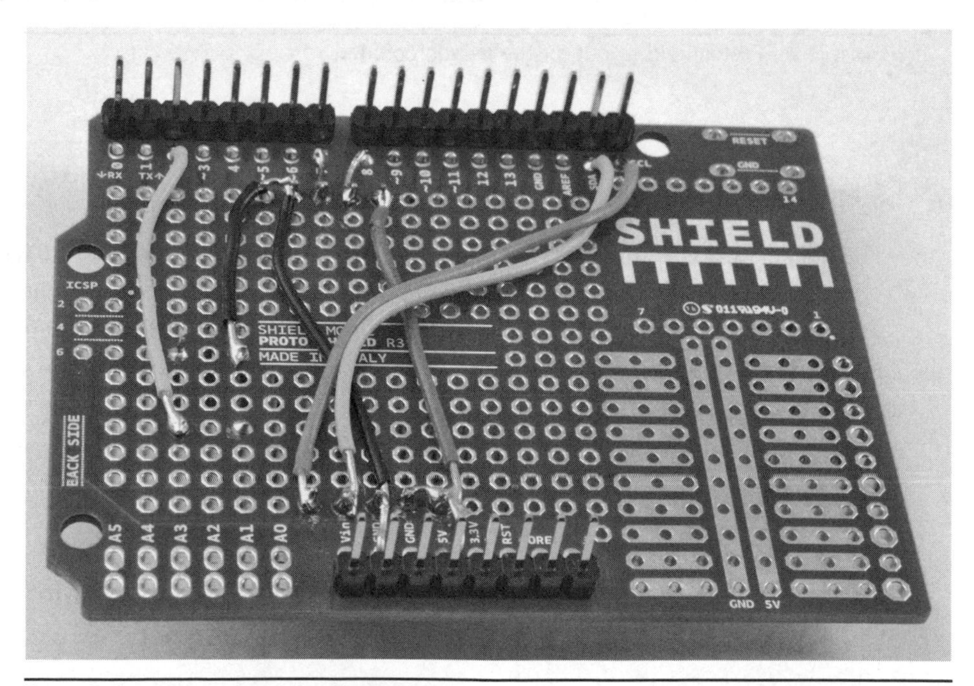

FIGURE 9-4 Underside of the Protoshield.

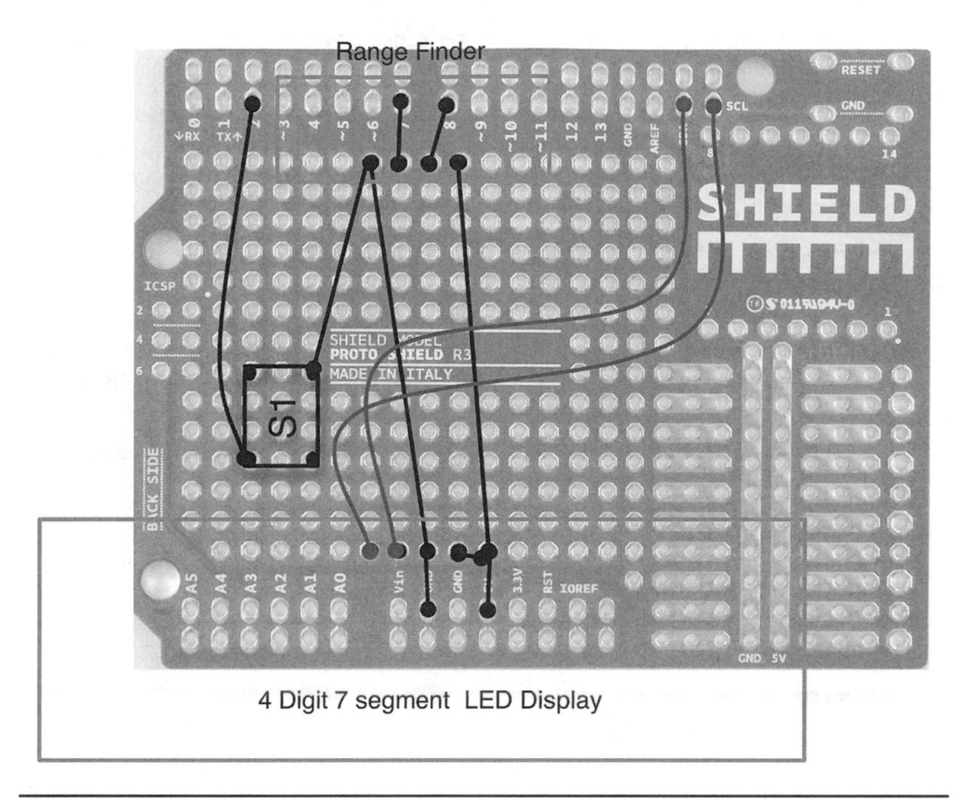

Figure 9-5 Wiring diagram for the underside of the board.

Software

To install the libraries for the LED module, go to the following URL and select the option to download as a zip file: https://github.com/adafruit/Adafruit-LED-Backpack-Library. The icon on the webpage is labeled "Zip" and looks like a cloud with an arrow coming out of it.

The zip file is small, so it will not take long to download. When it has downloaded, unzip the folder and rename it from "Adafruit-LED-Backpack-Library-master" to "AdafruitLEDBackpack." Then copy it to your "Arduino" folder.

The sketch for this project is ch_09_person_counter. After the library `include` statement, you will find the definition for the constant `threshold`. This specifies the maximum distance of an object from the sensor for that object to be counted. This distance is in centimeters. This is followed by the usual pin definitions. The ultrasonic range finder uses two Arduino pins, `trigger` and `echo`.

To count objects moving past the sensor, we need to detect when they are inside the threshold distance and then wait until they move further away than that distance. To keep track of these two states, we use the constants `waitingForArrival` and `waitingForDeparture` and the variable `state`.

```
int count = 0;
const int waitingForArrival = 0;
const int waitingForDeparture = 1;
int state = waitingForArrival;
```

The `count` variable is used to keep count of the number of objects that have passed the sensor.

The `setup` function sets the modes of the Arduino pins and initializes the display.

```
void setup()
{
  pinMode(buttonPin, INPUT_PULLUP);
  pinMode(trigPin, OUTPUT);
  pinMode(echoPin, INPUT);
  Wire.begin();
  display.begin(0x70);
}
```

The main `loop` function is as follows:

```
void loop()
{
  if (state == waitingForArrival && takeSounding() < threshold)
  {
    state = waitingForDeparture;
  }
  if (state == waitingForDeparture && takeSounding() >= threshold)
  {
    state = waitingForArrival;
    count ++;
  }
  if (digitalRead(buttonPin) == LOW)
  {
    count = 0;
  }
  display.print(count);
  display.writeDisplay();
```

```
  delay(50);
}
```

This `loop` function uses a state variable so that each time through the loop, it knows whether it should be waiting for an object to come into range (`state = waitingForArrival`) or it should be waiting for the object to leave (`state = waitingForDeparture`). It cannot literally wait for such a thing using a `while` loop because it needs to also keep checking for key presses.

If the `state` is currently `waitingForArrival` and an object has come close enough, then the `state` is changed to `waitingForDeparture`. The next time around the loop, the `state` will now be `waitingForDeparture`, and unless the object has departed, it will stay in this state. When the object does depart, the count will be incremented by one, and `state` will return to `waitingForDeparture`.

The next section of the `loop` function checks for a button press. If one occurs, then it sets the count back to 0. Finally, the count is written to the display, and the display is updated.

The function `takeSounding` sends a short pulse to the trigger pin of the ultrasonic rangefinder and then times how long it takes for the echo pin to go `HIGH`, indicating that the sound wave has returned to the module. The distance then can be calculated from this time taken.

```
int takeSounding()
{
  digitalWrite(trigPin, LOW);
  delayMicroseconds(2);
  digitalWrite(trigPin, HIGH);
  delayMicroseconds(10);
  digitalWrite(trigPin, LOW);
  delayMicroseconds(2);
  int duration = pulseIn(echoPin, HIGH);
  int distance = duration / 29 / 2;
  return distance;
}
```

Using the Project

You probably will need to adjust the direction in which you point the rangefinder and the value of `threshold` to get it to work well. Pressing the button will set the count back to zero.

There are some interesting uses for this kind of project. For example, if it were battery powered and waterproof, then you could use it to keep track of how many lengths of a swimming pool you had swum. In this case, you probably would want to increment `count` by two each time you got close to the edge of the pool where the counter is stationed.

Summary

This is a simple project that makes good use of ready-made modules. It is also one that lends itself to various adaptations. In Chapter 10 you will find the final security-related project in this book, in the form of a laser-based intruder alarm.

CHAPTER 10

Laser Alarm

Difficulty: ★★	Cost guide: $15

This project uses a laser module and a mirror to make a spy-styled laser alarm. In this case, the triggering of the alarm activates a relay. This offers the best flexibility in attaching sirens, lamps, or various other devices in response to the alarm being triggered. Once triggered, the alarm remains active until the reset button is pressed (Figure 10-1).

An alarm like this is really more for fun than for serious security. It is actually fairly easily defeated using a sheet of paper placed close to the laser and an LED so that the light from the laser is reflected straight back into the LED.

FIGURE 10-1 Laser alarm.

WARNING Lasers, even low-power lasers like the one used here, can seriously damage your eyes if you look directly into them. Because you are using the laser with a mirror, you need to be especially careful not to reflect it back into your eyes. To see if the laser is on, never look into it. Instead, hold a piece of paper in front of it to see the dot.

Parts

This project uses a Protoshield and a low-cost laser module. It also uses a clever trick to allow a large LED to be used as both the light sensor and an indicator of the alarm being triggered.

To build this project, you will need the following:

Part	Quantity	Description	Appendix
	1	Protoshield bare printed circuit board (PCB)	A3
	1	Laser module[a]	M12
	1	10-mm diffuse LED[b]	S8
	1	Header strip, male	H1
R1, R2	2	270 Ω, ¼ W resistor	R1
T1		Transistor 2N3904	S10
D1		Diode 1N4001	S11
Relay		5 V ("sugar cube") relay	C4
S1	1	Tactile push switch	C2
	1	Small mirror	
	1	Two-way screw terminal, 0.2 inch	H7

[a]Look for a low-power laser module; 1 to 5 mW is fine. Any more than that and you will need to use a transistor to control it. The module used was advertised as 5 V and for Arduino. It only consumes 20 mA and is therefore ideal for driving directly from an Arduino digital output.

[b]The LED should be large (10 mm ideal) to maker a bigger target when lining up the reflected laser beam. It also should have a diffuse lens and be the same color as the laser. I used a red LED and laser module, but green lasers have the cool feature of being far more visible in normal lighting conditions.

Protoshield Layout

Figure 10-2 shows the Protoshield layout for this project.

If you are happy with just the LED flashing when the alarm is triggered, then you can omit T1, the relay, D1, and R2. You can always add them later.

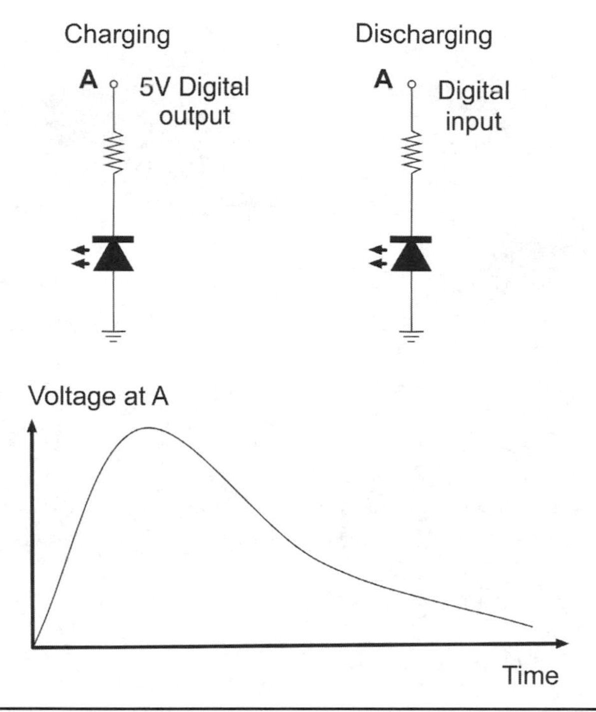

Figure 10-2 Protoshield layout for laser alarm.

Using an LED as a Light Sensor

LEDs can be used to make quite sensitive light detectors. Figure 10-3 shows how this works. Notice how the LED is reverse biased. That is, it is the wrong way around if we want it to emit light. This way around, the LED will act like a capacitor (think of a mini rechargeable battery). If we make point *A* positive, then the LED will charge up through the resistor *R*. This will happen very quickly, so almost immediately there will be a voltage across it of 5 V. If we disconnect point *A*, the voltage would gradually drop back down to zero. The speed at which it drops will depend on the amount of light falling on the LED.

Thus, if having charged the LED we then switch point *A* to a digital input instead of an output, we can see how long it takes for the voltage to fall from being high (over about 2.5 V) to low (2.5 V or below). This time will give us a measure of the light intensity.

If you build the project as described below, as well as running the main project sketch, you also can run the test program ch_10_led_sensor_test. This will measure the light intensity and write it to the serial monitor every half a second. You can find

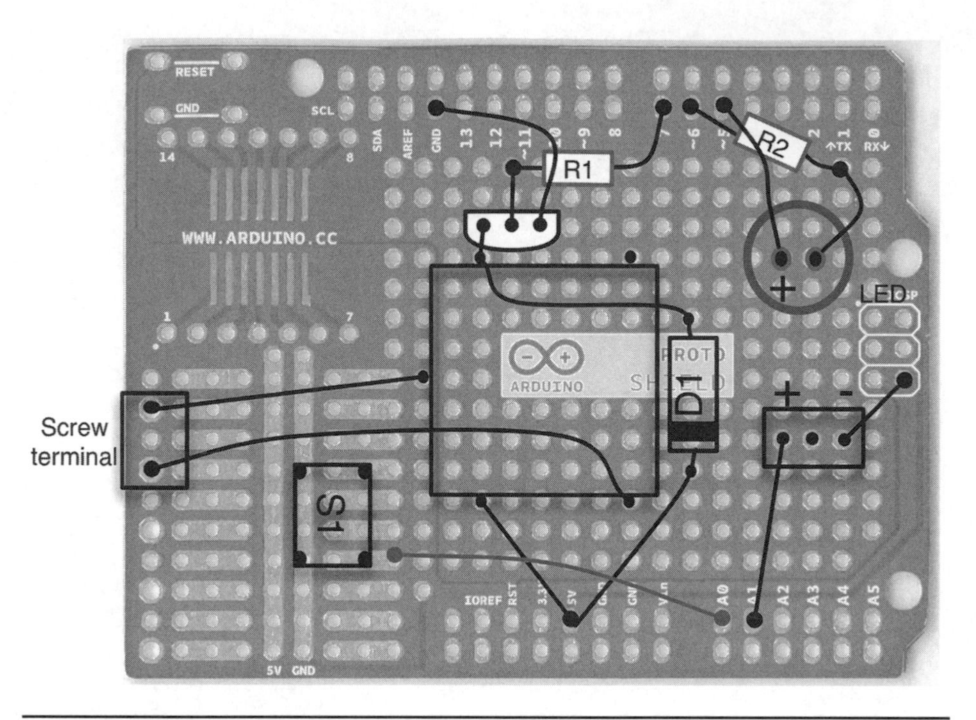

FIGURE 10-3 Using an LED as a light sensor.

an excellent description of this technique at www.thebox.myzen.co.uk/Workshop/LED_Sensing.html.

Construction

Aligning the Arduino and mirror so that the beam is neatly reflected back onto the LED is quite tricky and needs both the mirror and Arduino to be anchored securely in position. If you are going to use this project for real, then you probably also will want to run leads to the switch rather than have the switch on the board and easily pressable to cancel the alarm.

Figure 10-2 shows the Protoshield layout.

Step 1: Attach Header Pins to the Protoshield

Attach header pins to the Protoshield as done normally. Because stability is important for this project, it is probably a good idea to solder all the header pins.

Step 2: Solder the Relay onto the Protoshield

The type of "sugar cube" relay used in this project does not have pins in places that lend to their fitting through the holes on a Protosheild. Thus, the leads have been carefully bent out so that the relay can be soldered to the top surface of the board (Figure 10-4).

The relay now can be soldered to the top of the Protoshield in the position marked by Figure 10-4. Figure 10-5 shows the relay soldered into place.

Step 3: Solder the Remaining Components to the Protoshield

You can now solder the rest of the components to the board. Do not trim off the excess leads from the components because they will be useful later when you connect things up on the underside of the board.

There are a few things to check when you are soldering:

- The LED needs the longer positive lead to be toward the center of the board.
- Make sure that the transistor has its curved edge toward the relay.
- The stripe on the diode should be toward the bottom of the board.

Figure 10-6 shows the remaining components soldered into place.

Figure 10-4 Flattening the relay pins.

FIGURE 10-5 Soldering the relay.

FIGURE 10-6 All the components on the Protoshield.

Step 4: Solder the Underside of the Protoshield

You will be able to make most of the links on the underside of the board using the component leads. However, you will need a couple of longer lengths of wire for connections to the screw terminals and switch. Figure 10-7 shows the underside of the board, and Figure 10-8 shows the wiring diagram from the underside of the board.

Software

You can find the sketch for this project in ch_10_laser_alarm. The sketch does not require any Arduino libraries to be installed. The pin definitions for this sketch are as follows:

```
const int ledPinA = A0;
const int ledPinB = A1;
const int laserPin = A5;
const int buttonPin = 8;
const int alarmPin = 9;
```

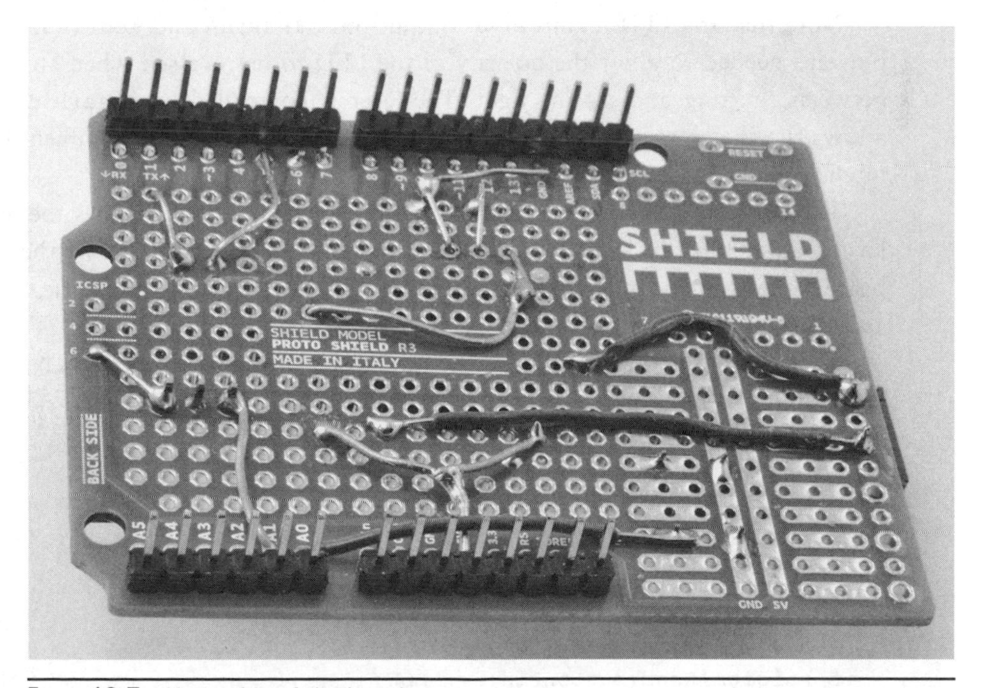

FIGURE 10-7 Underside of the board.

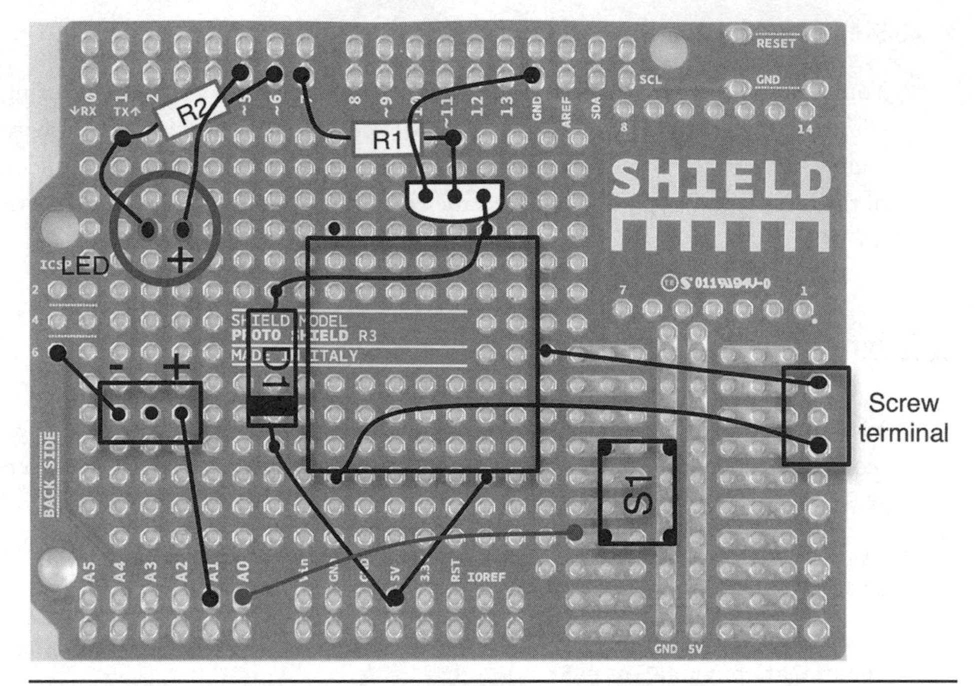

Figure 10-8 Wiring diagram for the underside.

Note that the LED requires two output pins, `ledPinA` and `ledPinB`. These pins are needed to allow the polarity of the LED to be reversed when switched between emitting and sensing light. The laser is controlled by `laserPin`. This allows the sketch to turn the laser off briefly, when measuring the ambient light intensity.

The constant `laserNoLaserRatio` determines by how many times the light-level reading must be lower with the laser lighted than without it before the laser beam is interrupted, and `alarmTriggered` is activated. A Boolean variable is used as a flag to indicate that the beam has been interrupted.

The `setup` function simply sets the pin modes of the various pins. The `loop` function is as follows:

```
void loop()
{
  if (! laserOnTarget())
  {
    alarmTriggered = true;
  }
  if (digitalRead(buttonPin) == LOW)
```

```
  {
    alarmTriggered = false;
  }
  digitalWrite(alarmPin, alarmTriggered);
  if (alarmTriggered)
  {
    flashLED();
  }
}
```

The first part of this function tests to see if the laser beam has been interrupted. If it has, then it sets `alarmTriggered` to `true`. It will remain set to `true` until the button is pressed. This is detected in the next section of `loop`. The `alarmPin` is then set to the same value as `alarmTriggered`. That is, if the alarm is triggered, the pin will be `HIGH`; otherwise, it will be `LOW`. Also, if the alarm is triggered, then the `flashLED` function will be called, temporarily reversing the polarity of the LED to make it light rather than act as a light sensor.

To decide whether the beam has been broken or not, the function `laserOnTarget` is used.

```
boolean laserOnTarget()
{
  long readingWithLaser = readLightIntensity();
  digitalWrite(laserPin, LOW);
  delay(10);
  long readingNoLaser = readLightIntensity();
  digitalWrite(laserPin, HIGH);
  return (readingWithLaser
    < (readingNoLaser / laserNoLaserRatio));
}
```

This function first reads the light intensity with the laser turned on, then turns the laser off, waits for 10 milliseconds, and then takes another reading, before turning the laser back on. The readings are lower the higher the light intensity, so the `laserOnTarget` function will only return `true` if the reading taken with the laser on is significantly less than the reading with no laser.

The following function reads the light intensity:

```
long readLightIntensity()
{
  // brighter = smaller number
```

```
  pinMode(ledPinA, OUTPUT);
  digitalWrite(ledPinA, HIGH);
  digitalWrite(ledPinB, LOW);
  delay(1);
  long startTime = micros();
  pinMode(ledPinA, INPUT);
  while (digitalRead(ledPinA)) {};
  long endTime = micros();
  return endTime - startTime;
}
```

It does this by first charging the reverse-biased LED by setting `ledPinA` to `HIGH` and `ledPinB` to `LOW`. It then delays for 1 millisecond before timing how long it takes for the voltage across the LED to drop until it is below the digital input on-off threshold. It then returns the time taken in microseconds. The lighter it is, the faster the LED will lose its charge.

You will remember that when the alarm is triggered, the LED is made to flash. This happens in the function `flashLED`.

```
void flashLED()
{
  pinMode(ledPinA, OUTPUT);
  digitalWrite(ledPinA, LOW);
  for (int i = 0; i < 5; i++)
  {
    digitalWrite(ledPinB, HIGH);
    delay(200);
    digitalWrite(ledPinB, LOW);
    delay(200);
  }
}
```

Before the LED can be used to light up rather than act as a light sensor, it must be forward biased by setting `ledPinA` to be an output and low. The pin `ledPinB` then can be used to flash the LED on and off in a loop.

Using the Project

In using this project, you will need to find a place to position the mirror so that it can reflect the laser light back onto the LED. The way to do this is to attach the mirror to a wall (self-adhesive putty is useful for this). The Arduino and shield then can be positioned opposite the mirror. Because there is a short distance between the laser and the LED, the laser module will need to be bent ever so slightly so that the reflected laser dot falls on the LED. You will see the LED glow when this happens.

A relay (Figure 10-9) is basically an electromagnet that closes switch contacts. The fact that the coil and the contacts are electrically isolated from one another makes relays great for things such as switching mains-powered devices on and off from an Arduino. Although the coil of a relay is often energized by between 5 and 12 V, the switch contacts can control high-power, high-voltage loads. For example, the relay shown in Figure 10-9 claims a maximum current of 10 A at 120 V alternating current (ac, mains) as well as 10 A at 24 V.

Because the relay contacts will behave just like a switch, you can use the relay to switch almost anything. The wiring diagram in Figure 10-10 shows how you could use the relay with a 12 V alarm sounder and power supply.

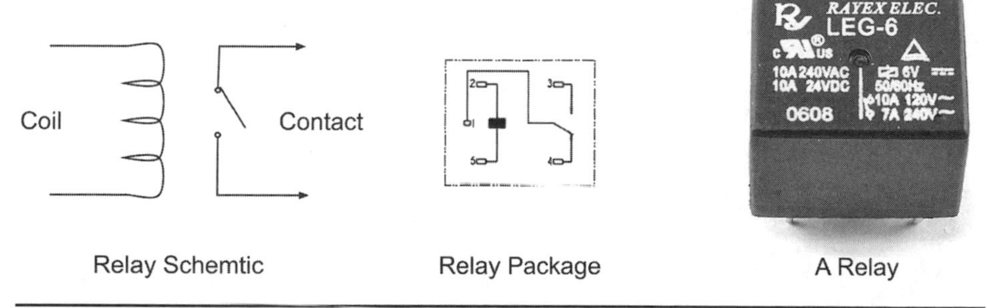

FIGURE 10-9 Relay.

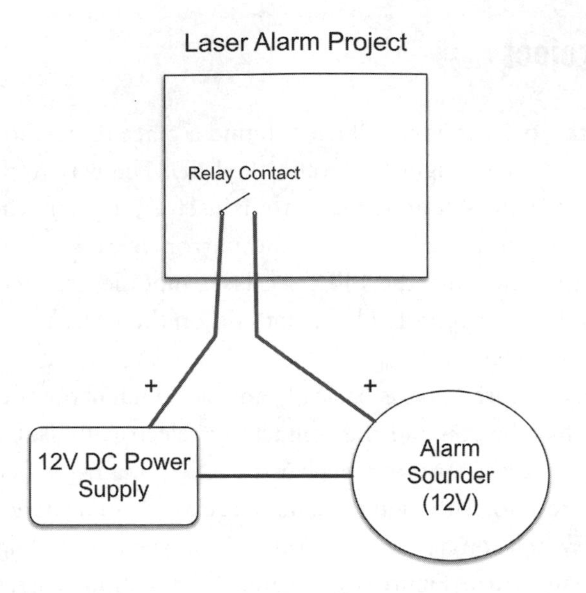

Figure 10-10 Using the project with a sounder.

Summary

This is the last of the security-related projects in this book. In the next section, you will find projects relating to sound and music.

PART THREE

Sound and Music

CHAPTER 11

Theremin-Like Instrument

Difficulty: ★★	Cost guide: $20

The theremin is a musical instrument developed in the 1920s by Russian Léon Theremin. The original instrument used radiofrequency electronics and a pair of antennas. By moving your hands in front of these antennas, you could control both the pitch and the volume that the instrument produced.

This Arduino version does not use radiofrequencies to convert hand positions into sound but rather uses an ultrasonic range finder. The distance sensed will determine the pitch of the sound (Figure 11-1).

Parts

To build this project, you will need the following:

Quantity	Description	Appendix
1	Protoshield printed circuit board (PCB) and header pins	A3, H1
1	Ultrasonic range finder HC-SR-04	M11
1	10 kΩ trimpot	R9
1	Powered PC speakers	
1	Strip of 0.1-inch header socket	H2
1	3.5-mm PCB socket	H4
1	1.5-mm drill bit	

Figure 11-1 Theremin-like instrument.

The ultrasonic range finder module was bought on eBay and has the name HC-SR-04. These devices are inexpensive and easy to find. You can, if you prefer, solder the range finder directly to the shield or, as described here, use a header socket.

Construction

This is a simple project to make. The only slightly tricky part is to solder the 3.5-mm headphone socket onto the shield. Figure 11-2 shows the schematic diagram for the project, and Figure 11-3 shows the layout of the components on the Protoshield.

Step 1: Solder the Header Pins to the Protoshield

Solder the header pins to the Protoshield. If you are unsure how to do this, refer to the Introduction to this book.

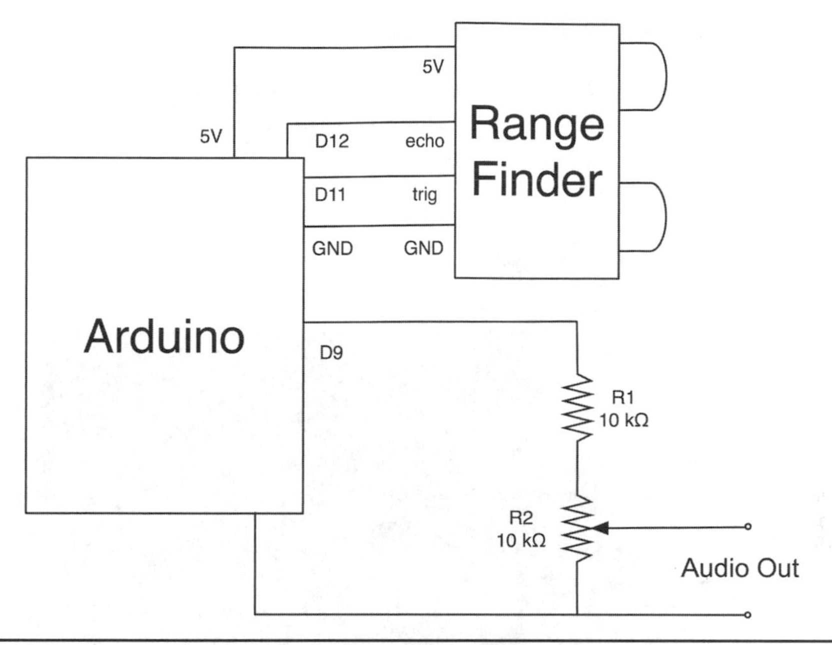

FIGURE 11-2 Schematic diagram of the instrument.

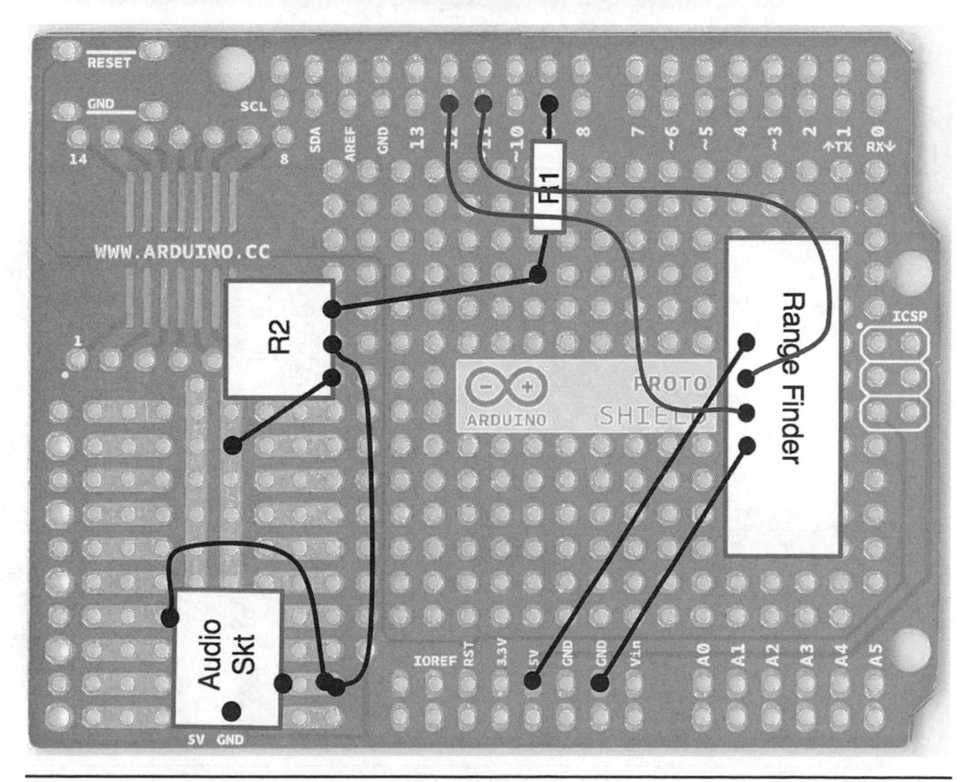

FIGURE 11-3 Protoshield layout

Step 2: Attach the 3.5-mm Socket

To attach the socket, you are going to splay out the leads on the side of the socket and drill the hole in the Protoshield under the center pin so that it will fit through (Figures 11-4 and 11-5).

Figure 11-4 Splaying the leads on the 3.5-mm socket.

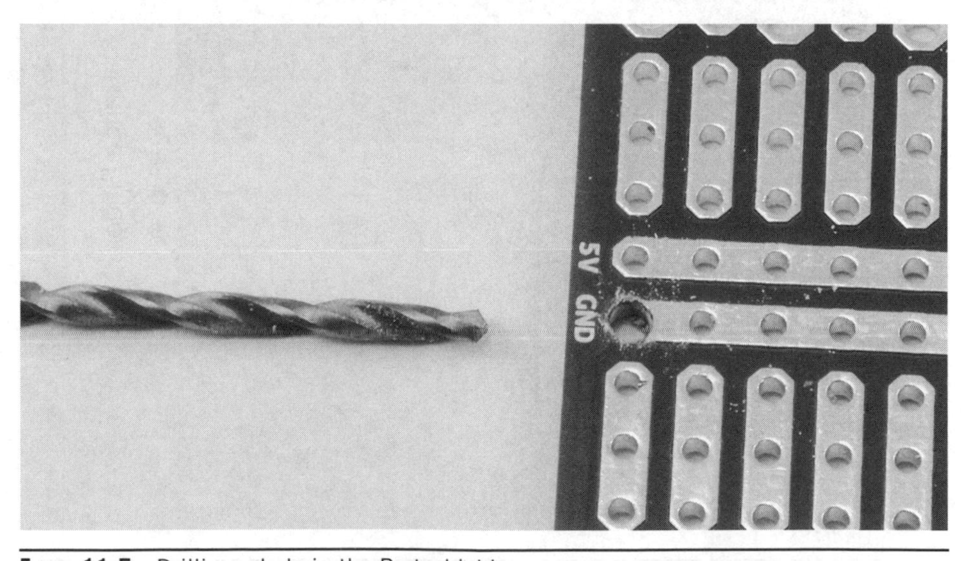

Figure 11-5 Drilling a hole in the Protoshield.

Figure 11-6 Protoshield with socket attached.

There is only one hole to drill, and because you are just widening the hole to 1.5 mm, you probably can drill it by just twisting the bit between your fingers. Figure 11-6 shows the Protoshield with the socket soldered in place.

Step 3: Solder the Remaining Components

The rest of the components are just soldered in the normal way. Cut off the excess lead from the resistor at the end nearest the Arduino header, but leave the other end long because you will use it to connect to R2 (the variable resistor). When all the components are in place, the top of the board will look like Figure 11-7.

Step 4: Link the Components

Start by bending the remaining resistor lead so that it joins one end of the variable resistor.

FIGURE 11-7 Top of the Protoshield.

You will need to use linking wires for all the remaining connections. Use Figure 11-3 as a guide for where the wires should go. Use insulated wires where they cross other wires.

When all the links are in place, the bottom of the board should look like Figure 11-8. Figure 11-9 shows the wiring layout from the underside of the board.

Software

This project uses a very sophisticated Arduino library to do all the clever tone-generation stuff. Therefore, before you can upload the sketch, you will need to get the library ("Mozzi") from http://sensorium.github.com/Mozzi/.

To install the library, select the download option that downloads the whole thing as a zip file, and save the directory into the "Libraries" folder inside your "Arduino" folder, which in turn will be in your "Documents" folder. The library does most of the work for us, leaving a pretty simple sketch, which can be found in ch_11_theremin.

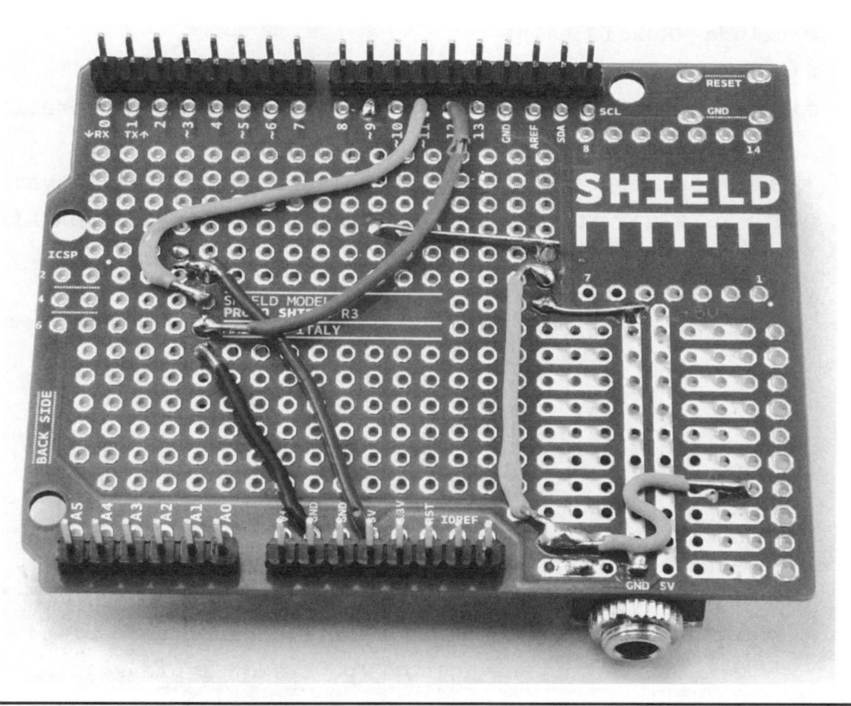

FIGURE 11-8 Completed Protoshield.

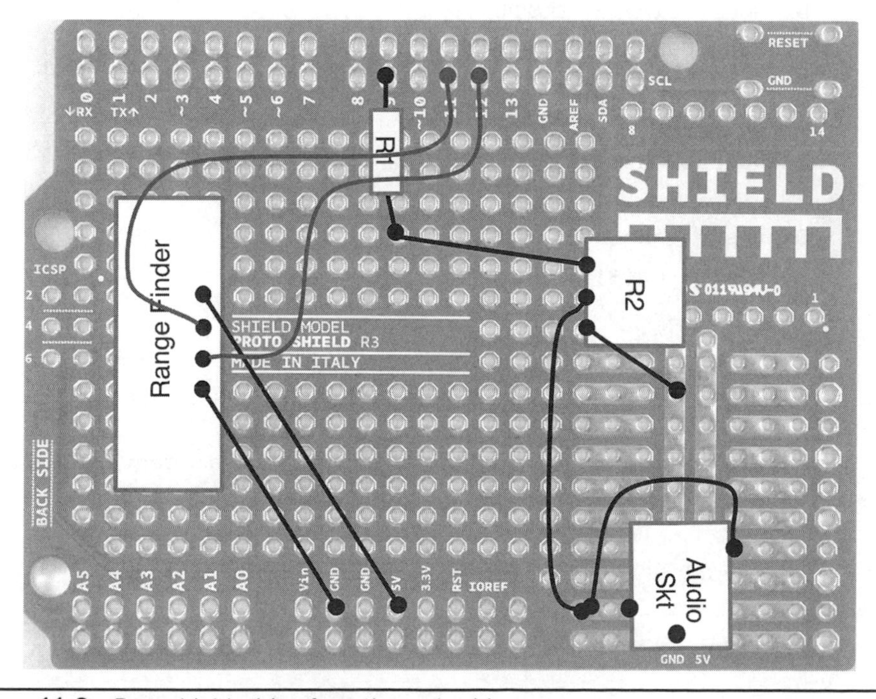

FIGURE 11-9 Protoshield wiring from the underside.

```
#include <MozziGuts.h>
#include <Oscil.h> // oscillator template
#include <tables/sin2048_int8.h> // sine table for oscillator

// use: Oscil <table_size, update_rate> oscilName (wavetable)
Oscil <SIN2048_NUM_CELLS, AUDIO_RATE> aSin(SIN2048_DATA);

// use #define for CONTROL_RATE, not a constant
#define CONTROL_RATE 64 // powers of 2 please

int trigPin = 11;
int echoPin = 12;


void setup()
{
  startMozzi(CONTROL_RATE); // set a control rate of 64
                            // (powers of 2 please)
  aSin.setFreq(440u); // set the frequency with an unsigned
                      // int or a float
  pinMode(trigPin, OUTPUT);
  pinMode(echoPin, INPUT);
}

void updateControl()
{
  float s = takeSounding();
  aSin.setFreq(s);
}

int updateAudio()
{
  return aSin.next(); // return an int signal centered around 0
}

void loop()
{
  audioHook(); // required here
}
```

```
float takeSounding()
{
  digitalWrite(trigPin, LOW);
  delayMicroseconds(2);
  digitalWrite(trigPin, HIGH);
  delayMicroseconds(10);
  digitalWrite(trigPin, LOW);
  delayMicroseconds(2);
  int duration = pulseIn(echoPin, HIGH, 1000);
  if (duration > 1000) duration = 1000;
  return (float)duration / 2.0;
}
```

The "Mozzi" library expects two functions to be defined, `updateControl` and `updateAudio`. The function `updateControl` will be called at a period determined by `CONTROL_RATE`. The function sets the frequency of the tone being generated.

The function `updateAudio` gets the next value for the sound wave. Thus, for the sine wave we are using here, this will be the next value from a table of sine waves.

The range finder works by measuring how long it takes for a sound pulse to reflect from an object. To do this, you have to send a pulse to its `trig` pin and then wait for the `echo` pin to indicate that the sound has returned.

Using the Instrument

Connect the Protoshield up to the amplified speakers or even your home stereo system. The instrument will not start making a tone until your hand is about a foot away from the range finder. Moving it toward the range finder will lower the tone; moving it away will raise it.

Summary

The quality of the sound produced is not bad, considering that it is being generated by a lowly Arduino. In Chapter 12 we will turn our attention to making an Arduino-controlled FM radio receiver.

CHAPTER 12

FM Radio Receiver

Difficulty: ★★	Cost guide: $30

This project uses a very low-cost FM radio module and an Adafruit LED display module to make a simple FM radio receiver (Figure 12-1).

Predefined stations are stored in the sketch, and a push-button switch is used to change stations. You also could easily modify the sketch to change

FIGURE 12-1 FM radio receiver.

channels using serial commands. This project uses an audio jack to output the received stereo signal. This will then need amplifying by connecting it to powered speakers.

Parts

To build this project, you will need the following:

Quantity	Description	Appendix
1	Protoshield printed circuit board (PCB) and header pins	A3, H1
1	TEA5767 radio module	eBay
1	TEA5767 breakout board	M15
1	Powered PC speakers	
1	Strip of 0.1-inch header socket (four way)	H2
1	3.5-mm PCB socket	H4
1	1.5-mm drill bit	
	Solid-core wire	

The TEA5767 radio modules have solder connectors that are 1 mm apart rather than the 0.1-inch spacing found on the Protoshield. This means that you cannot solder these modules directly to the Protoshield, but instead, you need to use an adapter PCB that will allow the module to be used with the Protoshield.

Construction

Figure 12-2 shows the Protoshield layout for this board. One feature of the design is that the TEA5767 module is powered from the 3.3 V supply of the Arduino rather than the 5 V supply. The module will work just fine at 5 V, but it was found that when the display module was attached, a great deal of interference was generated on the 5 V supply line. Using the 3.3 V line avoids this and gives an interference-free sound.

Step 1: Solder the Header Pins to the Protoshield

This project only requires the Arduino pins from the connector SCL to D8 and the power connector opposite it. Thus you can leave out the other pin headers if you wish.

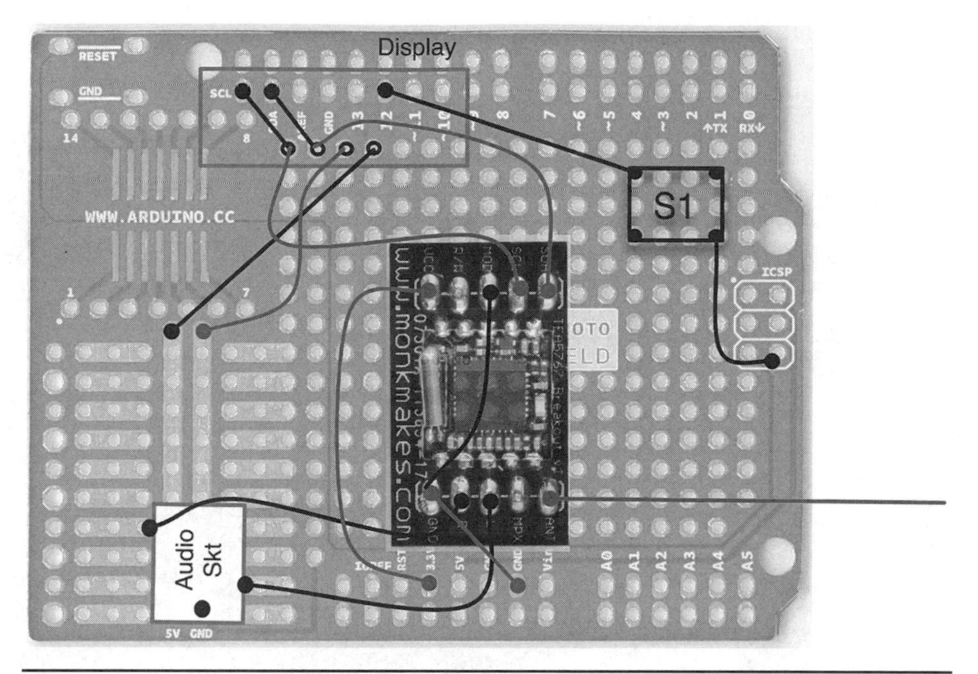

FIGURE 12-2 Protoshield layout for the FM radio.

Step 2: Assemble the TEA5767 Breakout PCB

You will find instructions for assembling this board on the MonkMakes website (www.monkmakes.com). Figure 12-3 shows the sequence of steps involved in this.

Figure 12-3*a* shows the module next to the PCB. Although it is tiny, the main problems when soldering it is not so much that it is small but that it is likely to move around. To fix this, a tiny piece of adhesive putty can be placed under the module. Squash the radio module onto this, making sure that it is oriented correctly with the metal tubular crystal at the bottom of the PCB. You can then carefully squish the module around a bit until its solder pads line up directly with the pads on the breakout PCB (Figure 12-3*b*).

You will need a soldering iron with a fairly fine point. Press it into the junction of the module connection and one of the pads on the PCB, and then feed in a bit of solder (Figure 12-3*c*). If you end up with solder bridging neighboring pads, you should be able to remove it easily with desoldering braid. Press the braid on top of the affected areas with the soldering iron.

Finally, solder on the header pins as you would with a Protoshield (Figure 12-3*d*).

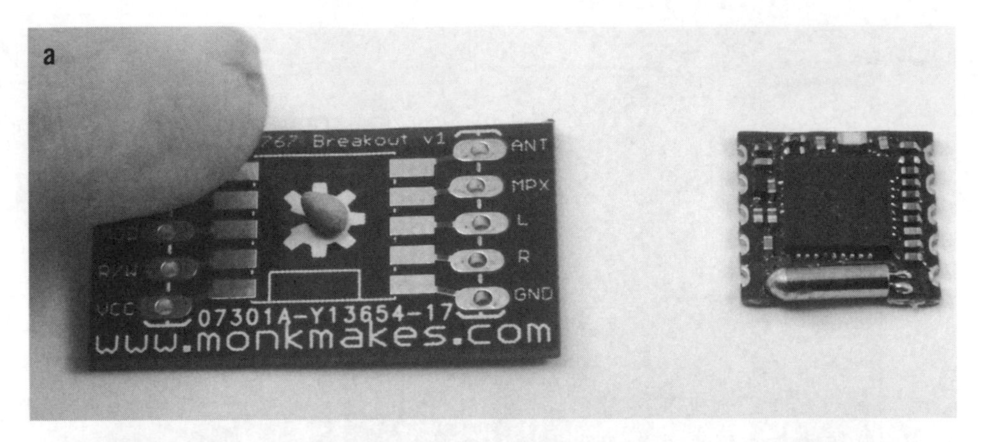

Figure 12-3 Assembling the TEA5767 breakout board.

Figure 12-3 Assembling the TEA5767 breakout board (*continued*).

Step 3: Attach the Audio Socket

This project uses the same audio socket in the same position as the project in Chapter 11. So please refer to that chapter to prepare the leads on the socket and drill out one of the holes so that it can fit securely onto the Protoshield. Then solder the socket to the board. When this is complete, it should look like Figure 12-4.

Step 4: Attach the Components

Now attach the switch, breakout board, and header socket to the Protoshield. Make sure that you get the breakout board the right way around (see Figure 12-2). Do not push the leads of the breakout PCB all the way through, or if you are going to use an Arduino Uno, the leads will foul the pins of the ATMega chip on the Arduino. Therefore, just push the pins through enough to solder them. This will leave the breakout PCB standing proud above the Protoshield by a few millimeters. When the components are all attached, the Protoshield will look like Figure 12-5.

Figure 12-4 Protoshield with socket attached.

Figure 12-5 Completed topside of the board.

Step 5: Solder the Underside of the Board

Figurc 12-6 shows thc wiring from thc undersidc of thc board. Notc that bccausc none of the modules have long leads, you will need a good supply of solid-core wire for hooking up the connections on the underside. When the underside of the Protosheild has been wired, it will look like Figure 12-7.

Step 6: Make an Antenna

The simplest kind of antenna to use with this receiver is a length of wire. This should be about 2 ft. long and can be either single or multicore wire. Strip one wire, and solder the other through the top of the board at a hold adjacent to the ANT pin of the TEA5767 module, and then connect the wire to that pin. Figure 12-8 shows this connection.

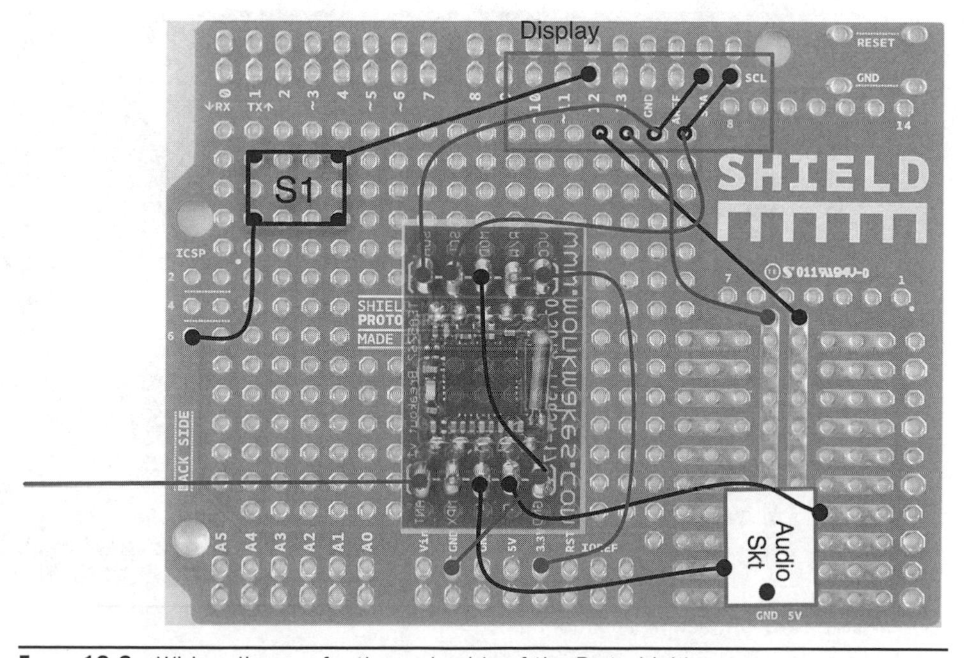

FIGURE 12-6 Wiring diagram for the underside of the Protoshield.

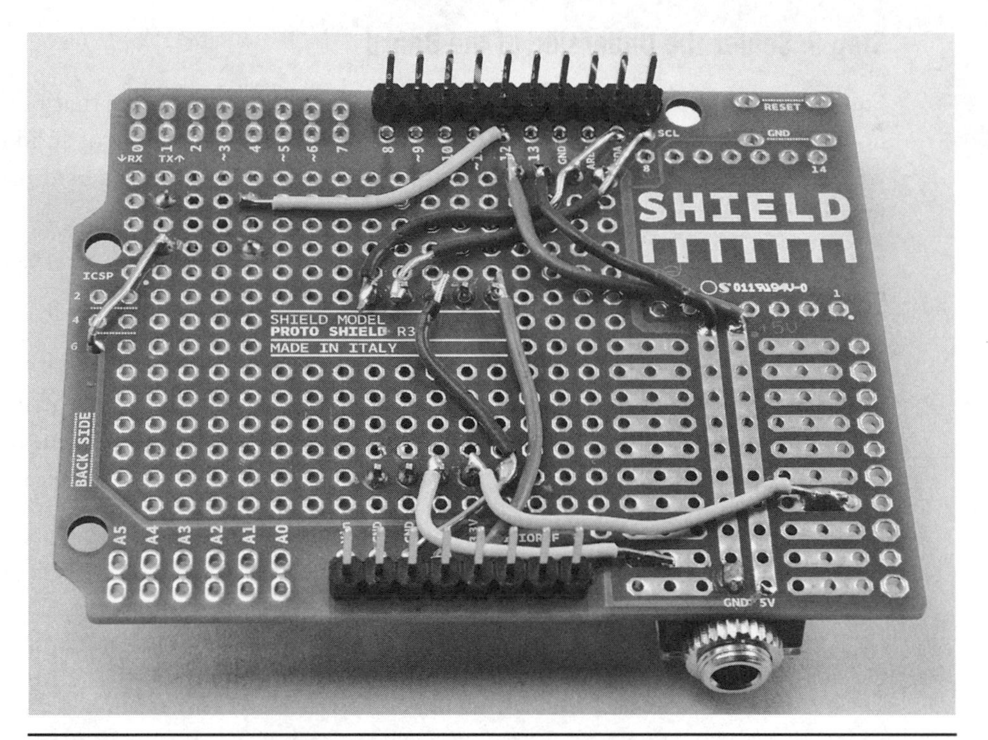

FIGURE 12-7 Underside of the Protoshield.

FIGURE 12-8 Attaching an antenna.

Software

The sketch for this project is titled ch_12_fm_radio. The project needs a few libraries to be installed into the Arduino environment:

- "Wire"—interintegrated circuit (I2C) interface
- "TEA5767Radio"—a library specifically for the TEA5767 module
- Adafruit display libraries

The "Wire" library is built into the Arduino IDE, but you will need to fetch the others. The Adafruit libraries are the same as those used in Chapters 15, 21, and 23. To install them, go to the URL (https://github.com/adafruit/Adafruit-LED-Backpack-Library) and select the option to download as a zip. The icon on the webpage is labeled zip and looks like a cloud with an arrow coming out of it.

The zip file is small, so it will not take long to download. When it has downloaded, unzip the folder and rename it from "Adafruit-LED-Backpack-Library-master" to "AdafruitLEDBackpack." Then copy it to your "Libraries" folder. You will need to do the same for the library "Adafruit_GFX" that you can find at https://github.com/adafruit/Adafruit-GFX-Library.

If this is the first time you have added a library, you will need to create a "Libraries" folder by opening the "Arduino" folder in your "Documents" folder and then creating a new folder called "Libraries." You can find the TEA5767 library at http://playground.arduino.cc/Main/TEA5767Radio, where there is a direct download of the zip file.

Before uploading the software, you will need to change the frequencies specified in the array `freqs`. Choose frequencies that you know correspond to radio stations in your area before uploading.

```
double freqs[12] = {98.3, 89.0, 91.0, 93.0, 97.4, 103.0, 0.0,
  0.0, 0.0, 0.0, 0.0};
```

The TEA5767 library is very easy to use. To start the radio, you just need the following line:

```
TEA5767Radio radio = TEA5767Radio();
```

A similar line is needed to start the display.

```
Adafruit_7segment display = Adafruit_7segment();
```

The setup function sets the buttonPin's mode to be INPUT_PULLUP, starts I2C communication, and initializes the display before displaying the current frequency using the displayFrequency function.

The loop function is as follows:

```
void loop()
{
  if (digitalRead(buttonPin) == LOW)
  {
    channel ++;
    if (channel == 12 || freqs[channel] == 0.0)
    {
      channel = 0;
    }
    radio.setFrequency(freqs[channel]);
    displayFrequency();
    delay(300); // debounce key
  }
}
```

This simply loops, doing nothing until the button is pressed. When this happens, the channel number is incremented. If this makes channel equal to 12, or if the freqs array value for that channel is 0, it sets the channel back to 0. The radio module is then told to change to the new frequency indicated in the array, and the display is updated.

Before returning to the loop, there is then a delay of 300 milliseconds to prevent key bouncing. Displaying the frequency requires a bit of manipulation of the frequency value because the Adafruit library does not directly support the display of floats. The function is listed as follows:

```
void displayFrequency()
{
  display.clear();
  float f = freqs[channel];
  int f10 = int(f * 10);
  int d4 = f10 % 10;
  f10 = f10 / 10;
  int d3 = f10 % 10;
  f10 = f10 / 10;
  int d2 = f10 % 10;
```

```
    f10 = f10 / 10;
    int d1 = f10 % 10;
    f10 = f10 / 10;
    if (d1 > 0)
    {
      display.writeDigitNum(0, d1, false);
    }
    display.writeDigitNum(1, d2, false);
    display.writeDigitNum(3, d3, true);
    display.writeDigitNum(4, d4, false);
    display.writeDisplay();
}
```

The first step is to clear the display, making it ready for the new value to be written. This is needed because the frequency may be either three or four digits, and we don't want the extra digit of the old displayed value appearing after the frequency has changed to three digits.

Next, a variable `f10` is defined that is an `int` value of 10 times the frequency. Thus, if the frequency for the current channel is 102.3, the value of `f10` will be 1,023. The lines of code that follow this split this four-digit number into four separate numbers by taking the modulo remainder and dividing the number by 10 and then dividing that number by 10 to move up through the digits.

The `if` statement is used to suppress the displaying of the first digit if it is 0. The display's `writeDigit` function takes three parameters. The first is the digit position. Rather confusingly, these are 0, 1, 3, and 4. That is, digit 2 is skipped because this digit is actually reserved for the colon in the center of the display. The second parameter to `writeDigit` is the single-digit value to display, and the final parameter is whether to display the dot for that digit. We are fixing the decimal place at the last but one digit, so this parameter is only true for digit 3.

Using the Project

When you press the button, the radio will cycle through the frequencies specified in the `freqs` array, so you will need to find the frequencies of your favorite radio stations and enter them in the sketch.

Summary

In the next couple of projects we will build devices for controlling music software such as Ableton Live. These are projects that use the Arduino Leonardo's keyboard emulation feature and can be easily adapted to other purposes.

CHAPTER 13

Pedal Board Controller

Difficulty: ★ / ★★★ (screw-shield version)	Cost guide: $5 / $20 (screw-shield version)

Although I have called this a *pedal board* for one version of the build, you will actually need pretty dainty toes to use it with your feet. I have, however, provided a second version of the design that is intended for use with separate switches that you might fix into a box and then run leads back to a screw shield on the Arduino.

The project is intended for use with music software such as Ableton Live as a simple controller. Therefore, you press one of the buttons, and that triggers some action within the music software—say, starting a drum track playing. In effect, you are making something that will emulate a computer keyboard that has most of the keys removed! Figure 13-1 shows the Protoshield version of the project, and Figure 13-2 shows the screw-shield version.

This is one project that does require the use of an Arduino Leonardo. This version of the Arduino has the ability to emulate a keyboard. This is something that is not possible with the Arduino Uno.

Parts (Protoshield Version)

This version of the project has tactile push switches attached to a Protoshield and is therefore of more use for fingers than feet.

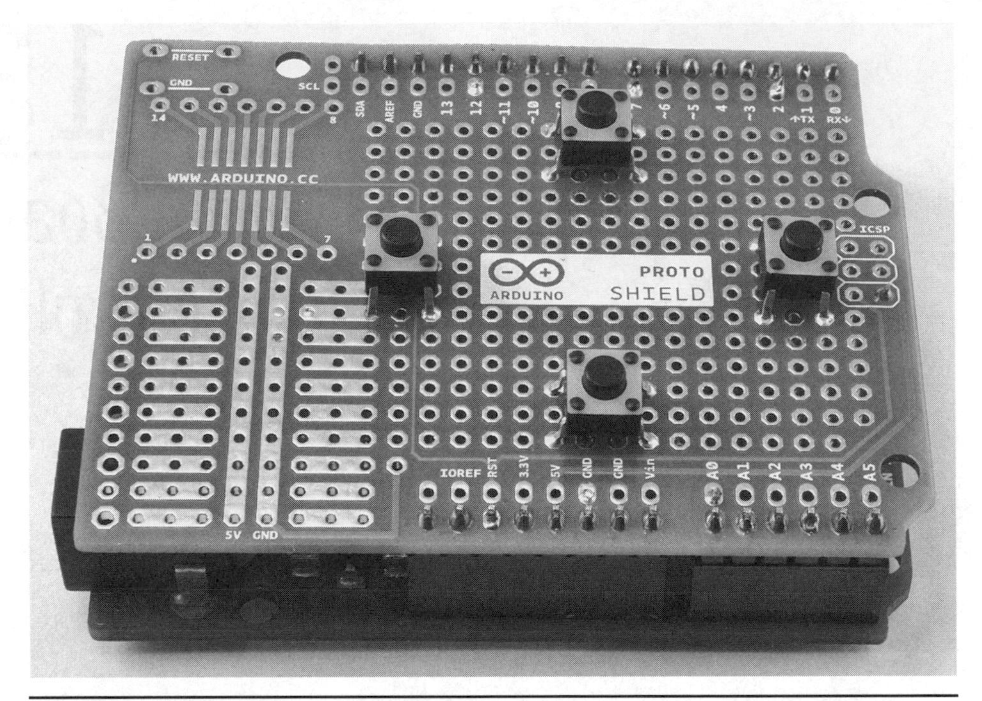

Figure 13-1 Pedal board controller (Protoshield).

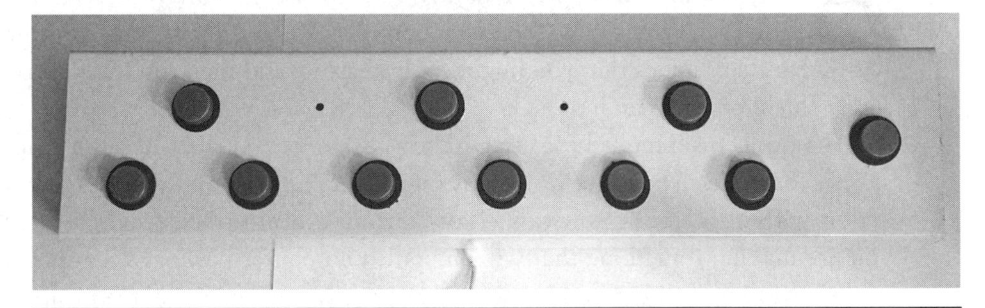

Figure 13-2 Pedal board controller (screw shield).

To build this project, you will need the following:

Part	Quantity	Description	Appendix
	1	Protoshield bare printed circuit board (PCB)	A3
	1	Header strip, male	H1
S1–4	4	Tactile push switch	C2

Protoshield Layout (Protoshield Version)

Figure 13-3 shows the Protoshield layout for this project.

Construction (Protoshield Version)

The switches are arranged in a diamond pattern around the Protoshield. Note that the switches have a pin spacing that is longer in one direction than in the other, and the switches are not all oriented the same way.

Step 1: Attach the Header Pins to the Protoshield

Attach the header pins to the Protoshield as normal. This project uses pins from all four headers, so you will need to solder on a full set of header pins.

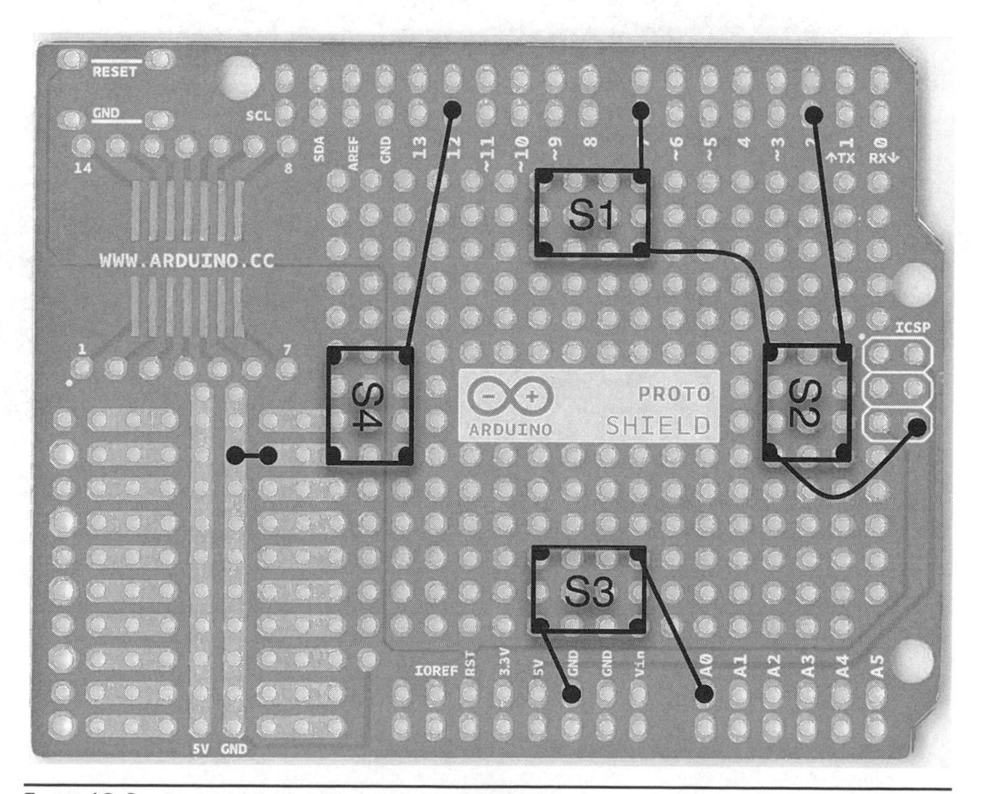

FIGURE 13-3 Protoshield layout for the pedal board.

Step 2: Solder the Switches to the Protoshield

Using Figure 13-3 as a reference and making sure that you have each switch the right way around, push the switches into the Protoshield. The switch leads can be a little delicate, so make sure that they are all lined up with the holes before pushing the switches firmly into place. Try to keep the switches level. When they are all soldered to the board, it will look something like Figure 13-4.

Step 3. Solder the Underside of the Protoshield

Figure 13-5 shows the underside of the board, and Figure 13-6 shows a schematic diagram of the underside. None of the wires cross over each other, so use Figure 13-6 as a reference to connect the components on the underside of the board.

Figure 13-4 Soldering the switches.

FIGURE 13-5 Underside of the Protoshield.

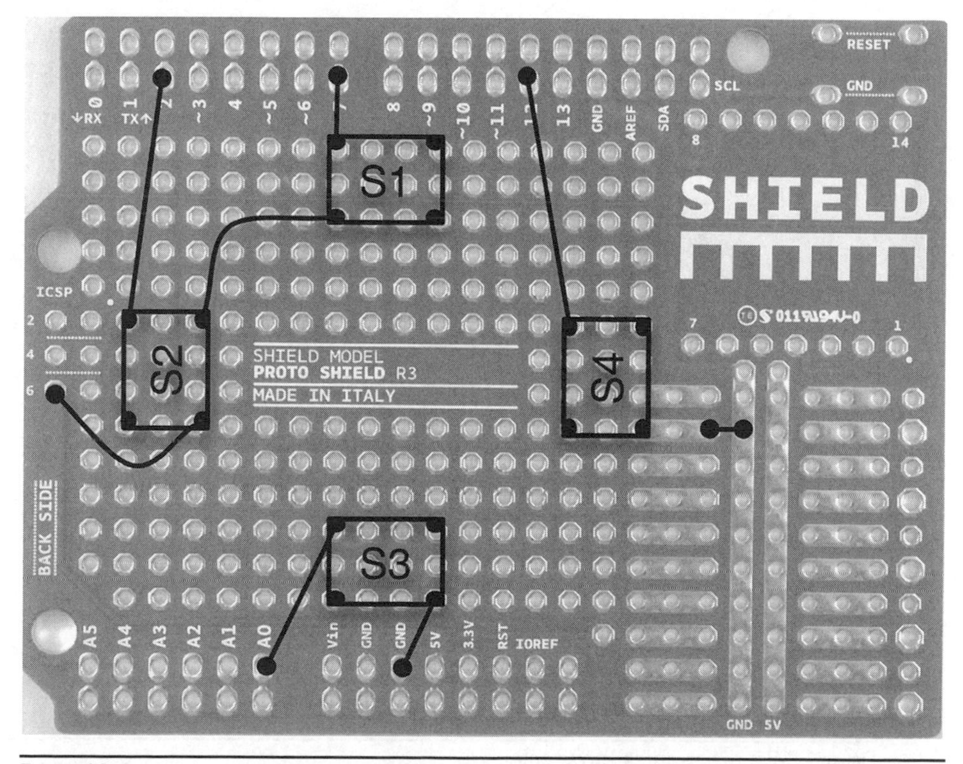

FIGURE 13-6 Wiring diagram for the underside of the board.

Parts (Screw-Shield Version)

This version of the project uses a ready-made shield called a screw shield (Figure 13-7). There are many versions of this kind of shield, and it allows you to attach wires to Arduino pins using screw terminals, which is perfect for, say, attaching a load of switches to an Arduino.

In this project, you can attach as many switches as you like to the Arduino, only limited by the number of pins available. Thus, you could attach pins to D0–D12 and A0–A5, which is 19 buttons. This design uses 10 buttons.

To build this project, you will need the following:

Part	Quantity	Description	Appendix
	1	Arduino screw shield	M13
S1–10	10	Push switches[a]	eBay
		Hookup wire	
		Enclosure	

[a]You should shop around for switches to suit the way you will use the project. You need any normally open push switch. These are available as durable foot switches, but they can be quite expensive.

FIGURE 13-7 Freetronics screw shield.

Construction (Screw-Shield Version)

This design is built into a section of rectangular pipe with access holes cut in one side to allow access to the switches and Arduino. You may prefer to find a more robust enclosure. All that is really necessary is that it has enough room to easily press the switches that you want to use.

Step 1: Prepare the Enclosure

Work out where the Arduino and screw shield are going to fit in the enclosure, and drill some holes to bolt the board securely into place. Also drill holes for the switches. Figure 13-8 shows the front of the prepared drain pipe from the front, and Figure 13-9 shows the rear.

Step 2: Fit the Switches

You can now fit the switches through the holes as shown. Line up the contacts on the back so that it will be easier to run the common GND wire that will go to one contact of every switch. With the switches in place, the enclosure looks like Figure 13-10.

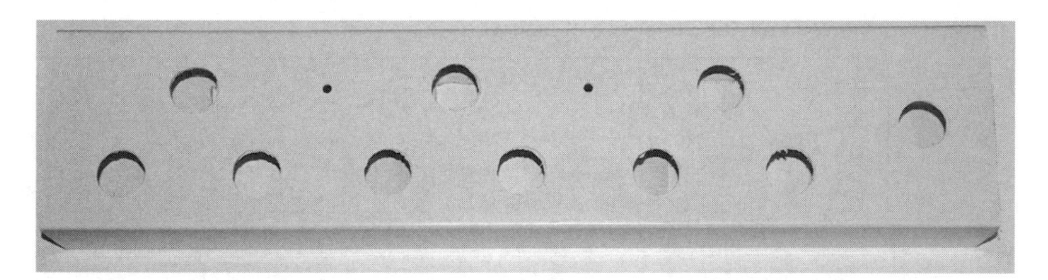

Figure 13-8 Preparing the enclosure (front).

Figure 13-9 Preparing the enclosure (rear).

Figure 13-10 Switches in place.

Step 3: Solder the GND Wires to the Switches

You can now solder wires to the switch connections and attach the other ends of the leads to the screw shield. Figure 13-11 shows a wiring diagram for the switches.

Because it is a little tricky to get access to the switches, I first connected together 10 short lengths of wire matching the GND connection to all the switches outside the enclosure (Figure 13-12) and then soldered them onto the switch terminals, leaving one lead to be connected to the screw terminal GND.

Step 4: Solder the Separate Wires to the Switches

The other terminals of the switches each need a longer wire reaching to it that will reach all the way to where you are positioning the Arduino and Protoshield. If the

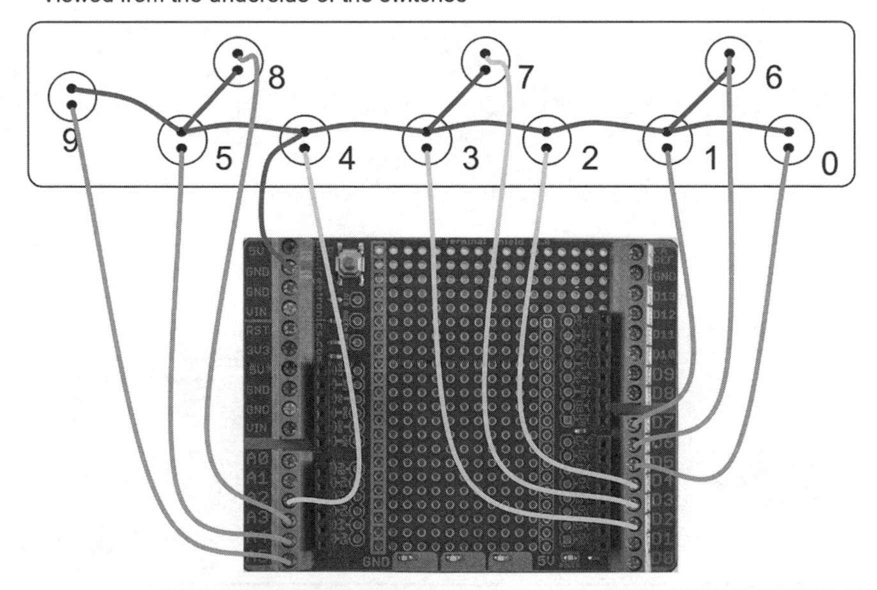

Figure 13-11 Wiring diagram of the switches.

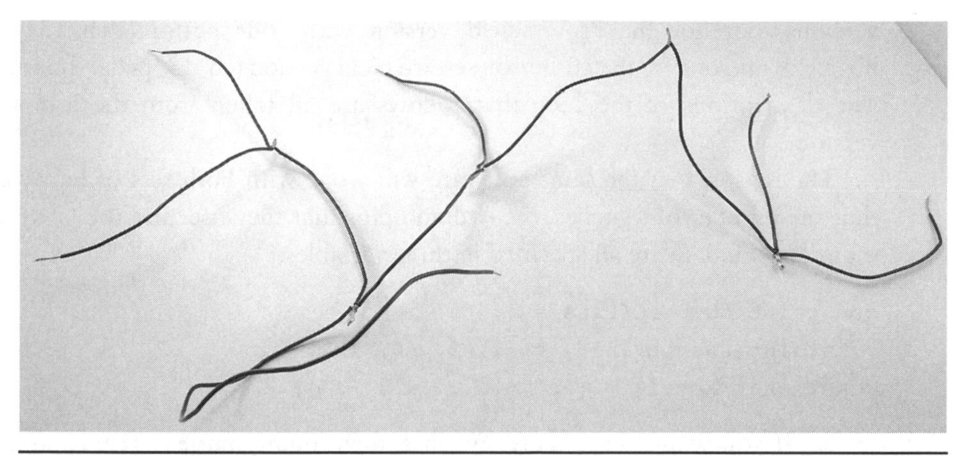

Figure 13-12 GND wiring.

wires are longer than they need to be, this does not really matter because they can just be tucked into the enclosure. It is certainly easier this way than with wires that are too short. In addition, it is probably best to attach each wire to the screw terminal as it is soldered to the switch, to avoid confusion about which wire is which. When all the wires are in place, the back of the pedal board should look something like Figure 13-13.

Software

The sketch is more or less the same, but you have many buttons and you must choose whether you use the Protoshield design or the screw-shield design. In the downloads for the book on my website (www.simonmonk.org), you will find two

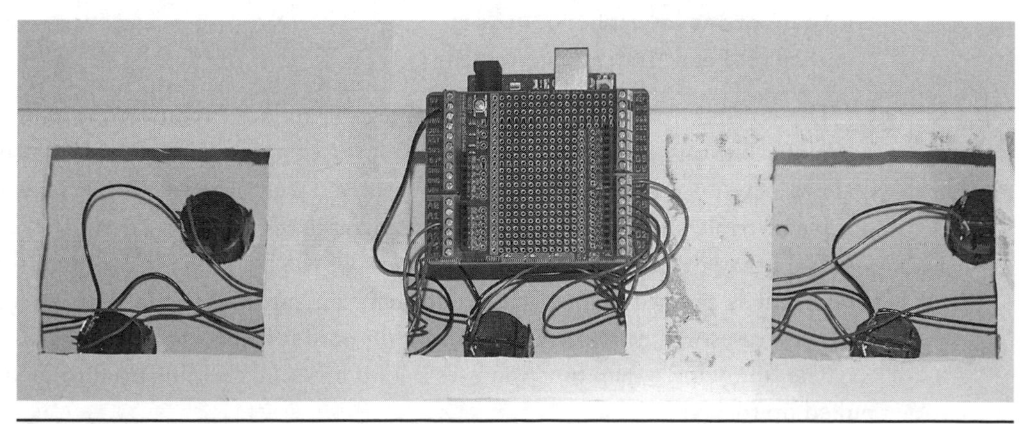

Figure 13-13 Completed wiring.

versions, one for the Prototshield version with four buttons (ch_13_pedal_board_4) and one for the 10-button screw-shield version (ch_13_pedal_board_10). The descriptions of the code that follows are all taken from the four-switch version.

The reason that the same software will work with both sets of hardware is that the number of switches, the Arduino pins that they use, and the key presses that they simulate are all specified in three variables.

```
const int numberOfPins = 4;
const int switchPins[] = {7, 2, A0, 12};
const char keys[] = {'a', 's', 'w', 'z'};
```

So, if you want to use more switches, then modify `numberOfPins` and then add other switch pins to the list of `switchPins` and a corresponding set of extra key names to the `keys` array. For an example of changing this to use 10 switches, see ch_13_pedal_board_10.

```
const int numberOfPins = 10;
const int switchPins[] = {5, 7, 4, 2, A2, A4, 6, 3, A3, A5 };
const char keys[] = {'0', '1', '2', '3', '4', '5', '6', '7',
   '8', '9'};
```

These are the only changes to make the sketch work with a different number and arrangement of switches.

The following section of code defines a constant and two more arrays used to make sure that the keys do not bounce, causing multiple triggerings for one key press, or *autorepeat*, sending the same key press continually while the key is depressed:

```
const long debouncePeriod = 50;
boolean pressed[numberOfPins];
long lastPressTime[numberOfPins];
```

The constant `debouncePeriod` specifies the time delay after one press of a particular button before that button can be pressed again. This is used later in the code to debounce the button presses. If you find that you are getting unwanted double or triple clicking of a button, then increase this value.

The array `pressed` is used to keep track of the state of each switch, that is, whether it is pressed or not. This is so that you can make sure that no further button presses will be registered for a button until it has been released.

The following `setup` function uses a loop to set all the pins used to be inputs pulled up to 5 V:

```
void setup()
{
  for (int i = 0; i < numberOfPins ; i++)
  {
    pinMode(switchPins[i], INPUT_PULLUP);
  }
  Keyboard.begin();
}
```

The `setup` function also puts the Arduino Leonardo into keyboard emulation mode using the `Keyboard.negin()` command.

Most of the work takes place in the following `loop` function:

```
void loop()
{
  for (int i = 0; i < numberOfPins ; i++)
  {
    boolean keyPressed = (digitalRead(switchPins[i]) == LOW);
    long timeNow = millis();
    boolean tooSoon = (timeNow < lastPressTime[i] +
      debouncePeriod);
    if (keyPressed)
    {
      if (pressed[i] == false && ! tooSoon)
      {
        Keyboard.print(keys[i]);
        lastPressTime[i] = timeNow;
        pressed[i] = true;
      }
    }
    else
    {
      pressed[i] = false;
    }
  }
}
```

The `loop` function will check all the switches listed in the `switchPins` array in turn. It first uses a `digitalRead` to see whether the switch is pressed and sets the local variable `keyPressed` accordingly.

It then sets a variable `timeNow` to the number of milliseconds since the Arduino last reset and then sets a second Boolean variable called `tooSoon` based on whether `debouncePeriod` milliseconds have elapsed since the key was last pressed.

We then have a pair of nested `if` statements. The outer one checks that the key was pressed. If it was, then the second `if` statement first makes sure that the key is not still pressed from last time and that sufficient time has elapsed since it was last pressed. If both these conditions are `true`, then it pretends to be a keyboard and sends the corresponding `key` for that switch. It then updates the `lastPressedTime` for that key and marks it as `pressed`.

In the situation where the switch was not pressed, its `pressed` status is set back to `false`.

You can test the program by opening a text editor or empty word processor document and then pressing the buttons to see the key commands appear.

Using the Project

One way of using this key pad is to make yourself a virtual drum set, where each button is allocated to a different drum sample. To do this in Ableton Live, the first step is to add four samples to the session view. You might, for example, use four different audio drum samples from your library arranged like the example of Figure 13-14.

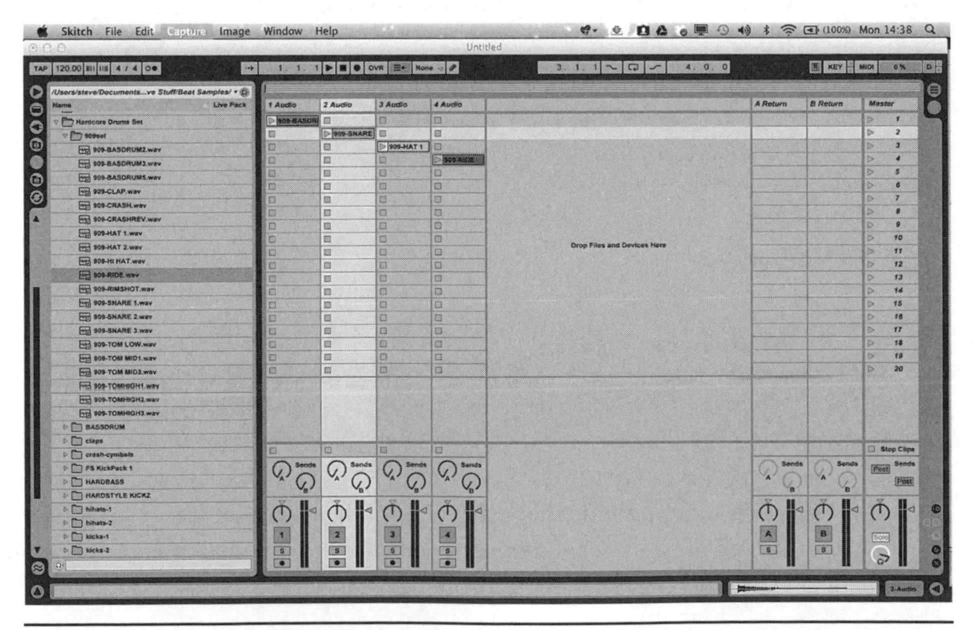

FIGURE 13-14 Samples in Ableton Live.

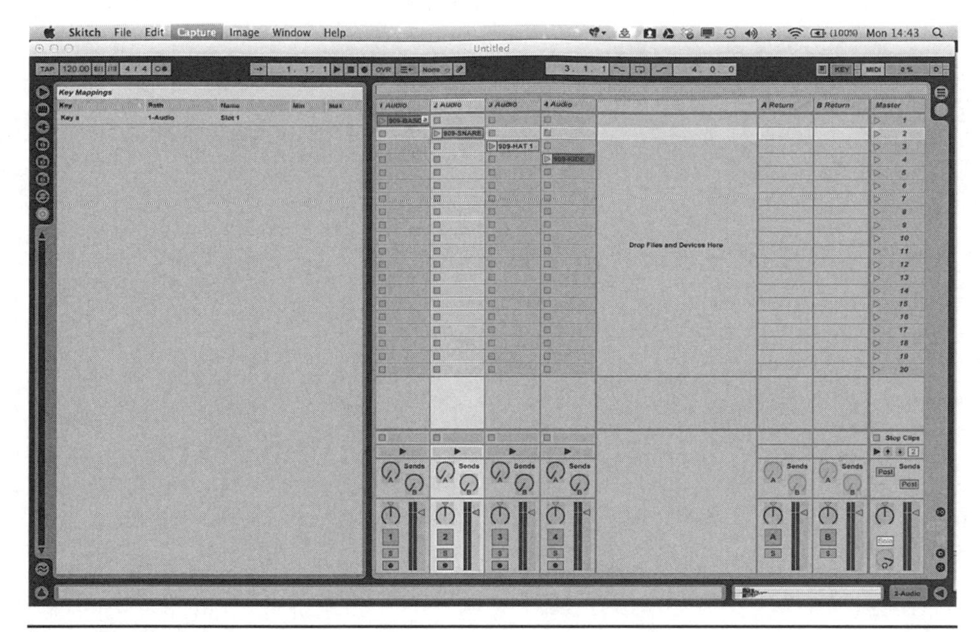

FIGURE 13-15 Mapping a key in Ableton Live.

The next step is to assign the keys to the different samples. To do this, first click on the "Key" button in the "Toolbar" area of Ableton Live. This puts Ableton Live into a key-mapping mode. Select one of the clips, and then press one of the buttons on the controller. You will see the key mapping appear in the "Key Mappings" area (Figure 13-15). Repeat this process for the other three buttons.

When you have finished mapping the keys, click on the "Key" button again to return to the session view. Before playing your virtual drum set, you will need to change the quantization to "None" from the quantization menu near the middle of the tool bar. You also could use this controller to switch between different drum beats or sections of a song to allow the basics of DJ-ing.

Summary

This is the kind of project that could be used for other things. You would have to change the sketch a little, but you could store passwords associated with each of the keys or other items of text that you might want to simulate typing at the touch of a button. The next project in this section is to build another music controller, this time one that generates key presses in response to movement by using an accelerometer module.

CHAPTER 14

Music Controller

Difficulty: ★	Cost guide: $20

This project is intended for use with music software such as Ableton Live as a simple controller. It uses an accelerometer module to detect movement and generate key presses (Figure 14-1).

An acceleration module can be used to measure the force of gravity acting on the module. You can use this effect to calculate the angle to which the module is tilted in both the *X* and *Y* directions. The module is a three-axis accelerometer that measures the force applied to a tiny weight inside the chip. Two of the dimensions (*X* and *Y*) are parallel to the module's printed circuit board (PCB). The third dimension (*Z*) is at 90 degrees to the module's surface. There normally will be a constant force acting on this dimension due to gravity. Therefore, if you tip the module, the effect of gravity starts to increase in the direction in which you tip it (Figure 14-2).

Parts

Just as with the preceding project, this project requires the use of an Arduino Leonardo for its ability to emulate a keyboard.

To build this project, you will need the following:

Part	Quantity	Description	Appendix
		Adafruit ADXL335 accelerometer module	M14

Figure 14-1 Music controller.

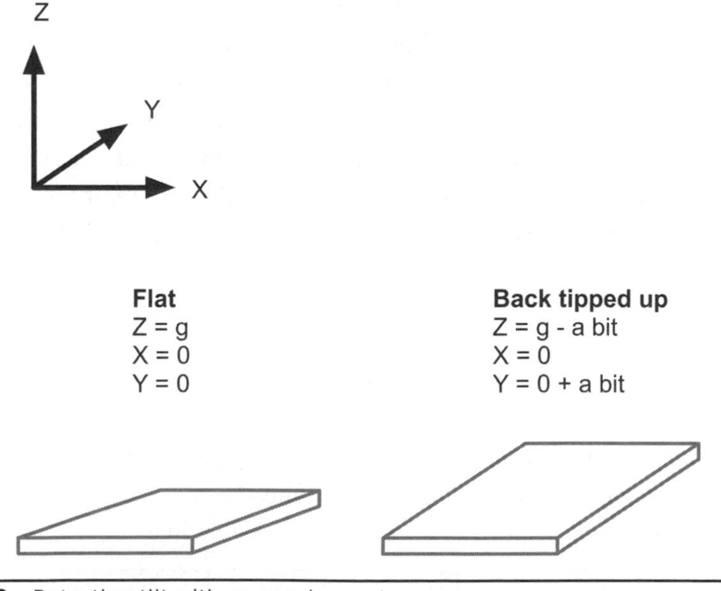

Figure 14-2 Detecting tilt with an accelerometer.

Construction

As you can see from Figure 14-1, this is a very simple project to build because the accelerometer module will plug directly into the analog socket section of the Arduino.

WARNING *Do not plug the accelerometer module into the Arduino until after you have successfully installed the sketch for this project. If pins A0–A5 have been used previously as digital outputs, then plugging the module in before uploading the sketch could easily damage the module, the Arduino, or both.*

Step 1: Solder Header Pins onto the Acceleration Module

The acceleration module is supplied without header pins attached. Thus, the first step is to solder them on. You will not be using the "Test" pin, so you can just solder the six pins from "Vin" to "Xout" (Figure 14-3). This will also then line up with the six pins A0–A5 on the analog connector.

FIGURE 14-3 Soldering header pins onto the accelerometer module.

Remember not to plug the accelerometer into the Arduino Leonardo before you have uploaded the sketch.

Software

The sketch for this project is titled ch_14_music_controller. The first constant in the sketch defines the keys to be used when the accelerometer is tilted back and forth. These map onto the keys used by Ableton Live's virtual keyboard.

```
const char keyNotes[] = "asdfghjk";
```

The next set of constants defines the pins that are to be used.

```
const int gndPin = A2;
const int xPin = 5;
const int yPin = 4;
const int zPin = 3;
const int plusPin = A0;
const int pin3V = A1;
```

To keep the wiring as simple as possible, power is supplied to the module through A0 acting as a digital output set to `HIGH`. The GND supply is provided by A2, which is also a digital output set `LOW`. The pins `xPin`, `yPin`, and `zPin` are used as analog inputs to read the forces acting in the *X*, *Y*, and *Z* directions.

The pin marked "3V3" on the accelerometer module is actually a regulated output voltage from the module. We do not want to use this, and we prevent it having any ill effects by setting the pin so that it is connected to a digital input.

The `offset` constant is the analog reading for the *X* and *Y* directions when the board is level. The setting of pin modes happens in the `setup` function as usual.

```
void setup()
{
  pinMode(gndPin, OUTPUT);
  digitalWrite(gndPin, LOW);
  pinMode(plusPin, OUTPUT);
  digitalWrite(plusPin, HIGH);
  pinMode(pin3V, INPUT); // 3V output - careful!
  Keyboard.begin();
}
```

The `setup` function also starts the keyboard emulation.

The `loop` function is as follows:

```
void loop()
{
  static int lastKeyIndex = 0;
  int x = analogRead(xPin) - offset;
  int y = analogRead(yPin) - offset;
  // -50 to + 50
  int keyIndex = (y + 50) / 12;
  if (keyIndex < 0) keyIndex = 0;
  if (keyIndex > 7) keyIndex = 7;
  if (keyIndex != lastKeyIndex && (abs(x) < 20))
  {
    Keyboard.releaseAll();
    Keyboard.press(keyNotes[keyIndex]);
    lastKeyIndex = keyIndex;
    delay(30);
  }
}
```

The `loop` function first reads the values of *X* and *Y* and adjusts them by subtracting offset. This gives a number that is more or less zero until the module is tilted.

Each time around the loop, the variable `keyIndex` is set to a value between `0` and `7` based on the value of `y`. The two single-line `if` statements just constrain the value of `keyIndex`.

The key is only simulated as being pressed if it has changed and the absolute value of `x` is less than 20; that is, the accelerometer is almost level in the *X* axis. Before the key is depressed, though, the `Keyboard.realeaseAll` command is used to release the previous key.

You can test the program by opening a text editor or empty word processor document and then tilting the board to see the key commands appcar.

Using the Project

To use this project with Ableton Live, the *Y*-axis movements will generate key presses corresponding to the keys on the virtual keyboard. Thus, all that you need to do is to drag a virtual instrument onto the session view, and make sure that the track for the virtual instrument is armed (the dot button at the bottom of the track

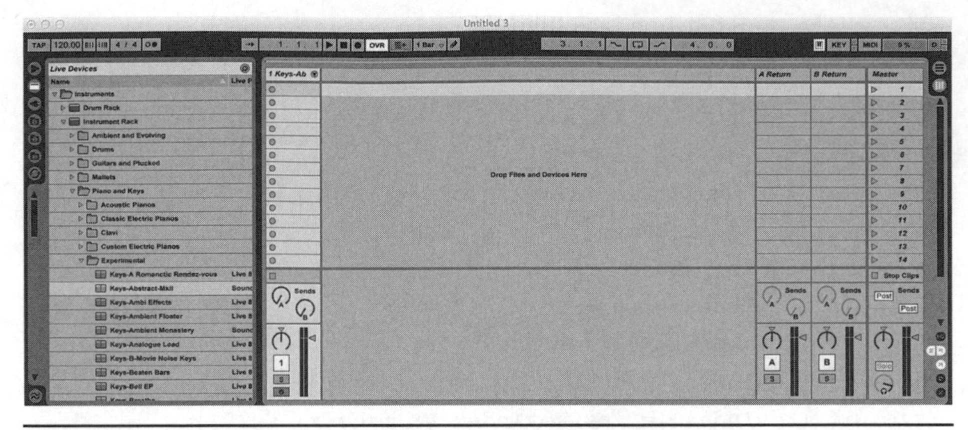

FIGURE 14-4 Using Ableton Live with the accelerometer controller.

containing the virtual instrument). Finally, turn on the computer MIDI keyboard using the icon on the toolbar. When this is fully set up, your Ableton Live window should look something like Figure 14-4. Tilting the controller forward and backward will turn the sound on and off, and tipping it slightly to the left or right will select the note to be played.

Summary

The next project is the final project in this section and uses a multicolor display to show a frequency spectrum of music or other sound used as its input.

CHAPTER 15
Spectrum Display

Difficulty: ★★★	Cost guide: $40

This project uses an Arduino to control a multicolor LED matrix, displaying the real-time frequency spectrum of an audio signal (Figure 15-1).

This project would make a nice addition to any audio system. Displaying the spectrum of an audio signal involves splitting the signal into a

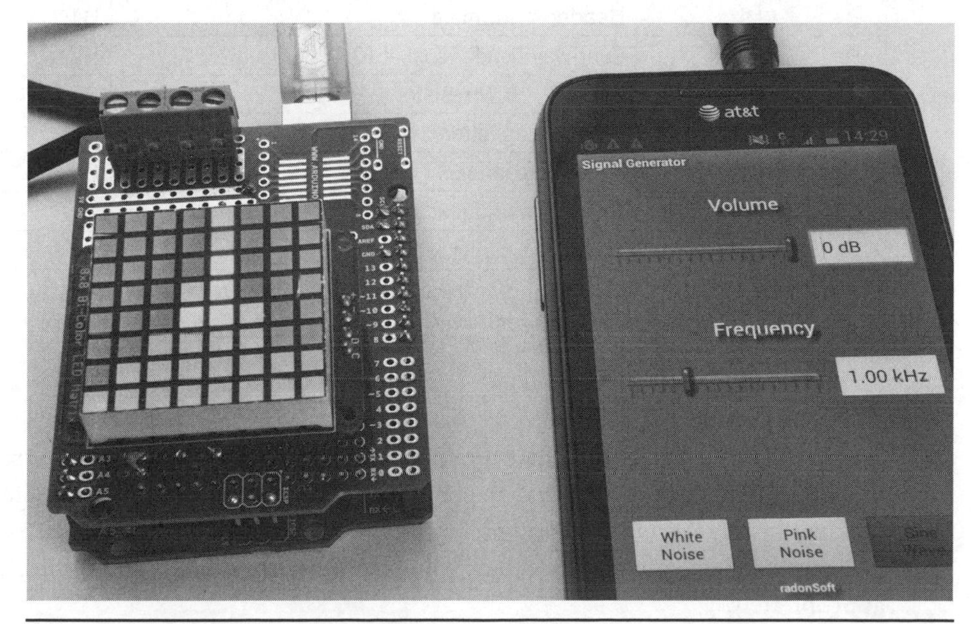

Figure 15-1 Audio spectrum display.

number of frequency bands and measuring the magnitude of the signal at each band. The human ear (or, rather, the young human ear) is usually assumed to be able to detect sound waves between 20 Hz and 20 kHz. Perception of pitch is logarithmic in nature, so the frequency bands are not evenly spaced.

You can also see in Figure 15-1 that the project is attached to an Android smartphone. This phone is running a free app called *Signal Generator*. This app can be used to generate a pure sine wave of some particular frequency. In this case, the frequency is 1 kHz, resulting in the peak frequency in the 1 kHz column of the LED display. Using such an app is a great way of testing the project before you try it on music. Such apps are also available for Apple and Windows smartphones as well as PCs.

Parts

To build this project, you will need the following parts:

Part	Quantity	Description	Appendix
	1	Adafruit LED matrix display	M4
IC1	1	MSGEQ7	S12
	1	8-pin DIL integrated circuit (IC) holder[a]	H10
	1	Protoshield bare printed circuit board (PCB)	A3
	1	Header strip, male	H1
	1	Four-way header socket (0.1 inch)	H2
R1	1	200 kΩ, ¼ W resistor	R5
C1–3	3	100 nF capacitors	C5
C4	1	33 pF capacitor	C6
	1	3.5-mm audio lead with plugs on both ends	
	1	Three-way screw terminal[b]	H13

[a]*The IC socket is optional. You can, if you prefer, solder the IC directly to the Protoshield.*

[b]*If you cannot find a 3-pin screw terminal, then one four-pin or two two-pin screw terminals are just fine. Simply leave the extra pin unconnected.*

To connect an audio device to this project, a normal 3.5-mm audio lead is sacrificed. It is cut in half so that the spectrum display sits between the source of the sound (say, an MP3 player or computer) and the amplifier responsible for making the signal audible. If you prefer, you can use a pair of 3.5-mm sockets like the one used in Chapter 11, but you will need to modify the Protoshield appropriately.

Sparkfun sells a shield using this chip, which also could be easily modified to add a socket to connect to the LED matrix display that would make a good alternative design. This shield actually uses two MSGEQ7 chips, allowing both the left and right audio signals to be monitored.

Construction

This project is relatively easy to put together. The screw terminal is used for both the incoming and outgoing audio signals, but only one of the two stereo signals is actually monitored for the spectrum display. Figure 15-2 shows the Protoshield layout for the project.

The MSGEQ7 Integrated Circuit

This project relies very heavily on the MSGEQ7 chip. This chip contains an amplifier and seven *notch* filters that are used to find the strength of the signal at seven different frequency bands centered on 63, 160, and 400 Hz and 1, 2.5, 6.25, and 16 kHz.

The chip does not have seven separate outputs but only requires one analog input and two digital outputs to interface with the Arduino. The analog output from the MSGEQ7 will be for one of the channels, and you can switch the frequency-band value that appears on this output using two control pins. When you send a pulse to the reset pin, it resets the chip so that it is ready to start taking readings. If you then send a pulse to the strobe pin, it will set the analog output to be read from the first frequency band.

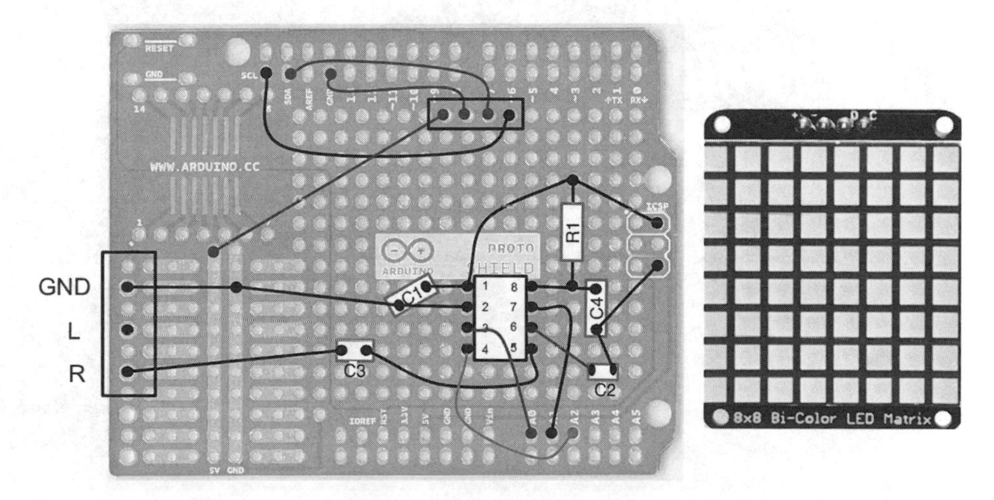

Figure 15-2 Protoshield layout for the spectrum display.

Send another pulse to the strobe pin, and it will move on to the next channel and so on for all seven filter values.

You can find out all about this chip from its datasheet, which is available at www.sparkfun.com/datasheets/Components/General/MSGEQ7.pdf.

Step 1: Solder the Header Pins onto the Protoshield

The Protoshield uses connections on only three of the four connection headers, so if you want to save some headers, just connect the two sections of the header pins that are used, that is, SCL to D8 and both connectors on the opposite side of the board.

Step 2: Solder the Resistor, Capacitors, and Integrated Circuit Holder

Start with the resistor (the lowest component on the board). Solder in the resistor, then the IC socket (or the IC if you are not using the socket), and then the capacitors. The IC and IC socket both have a notch indicating pin 1 of the IC. Make sure that this is toward the top of the board. Figure 15-3 shows the Protoshield with these components in place.

Figure 15-3 Protoshield with resistor, capacitors, and IC socket in place.

Do not snip off the excess leads of the resistor and capacitors yet. You can use them later to connect the underside of the board.

Step 3: Solder the Screw Terminals and Header Pin Socket

The header socket is supplied as long strips, and you will need to cut it to a length of four header sockets. To do this, score the strip on the fifth socket with a kraft knife, and then break the strip off over the edge of a desk. This means that you will sacrifice one of the sockets from the strip, but this works much better than trying to cut on the very narrow gap between the headers.

Solder the header socket onto the board. To get it straight, it is best to solder just one of the pins and then melt the solder and straighten up the header before soldering the other three pins. Another trick is to put a lump of adhesive putty around the header socket to keep it in place while you solder it.

Figure 15-4 shows the Protoshield with the remaining components attached. Note that I have used a four-way screw terminal block because I did not have a three-way block.

FIGURE 15-4 Protoshield with all components soldered into place.

Step 4: Connect the Underside of the Board

Use Figure 15-5 as a guide to wiring the underside of the board and Figure 15-6 for the finished wiring.

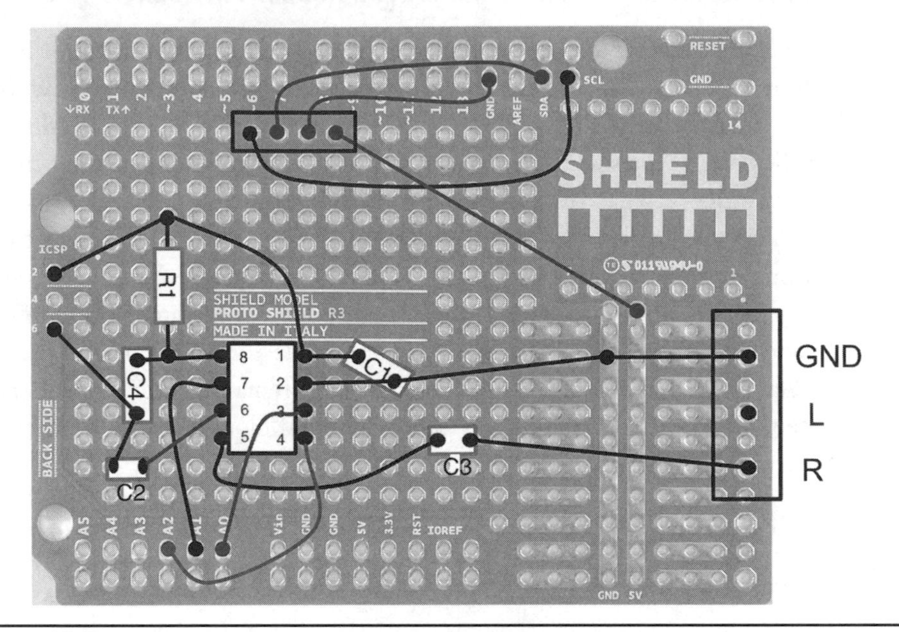

Figure 15-5 Wiring diagram for the underside of the Protoshield.

Figure 15-6 Underside of the Protoshield.

Start by using the existing leads on the components where possible. Where you have one wire that needs to cross another, use insulated solid-core wire.

Step 5: Fit the Integrated Circuit and Display

Integrated circuits are generally supplied with their leads a bit too far apart to easily fit into the IC socket, so you probably will need to bend the leads in a little before the IC will fit. The way to do this is to firmly hold the IC between finger and thumb and press one side of the leads down against your work desk (Figure 15-7). Be careful not to bend the leads too much. The IC must be inserted with the little notch in the package toward the top of the board. You can see the IC in place in Figure 15-8.

With the IC inserted, the board now can be fitted onto the Arduino while we prepare the audio lead.

Step 6: Prepare the 3.5-mm Audio Lead

Cut the audio lead in half, and strip the leads of both halves. The leads may have one outer shielded wire wrapped round a pair of inner insulated wires for the left

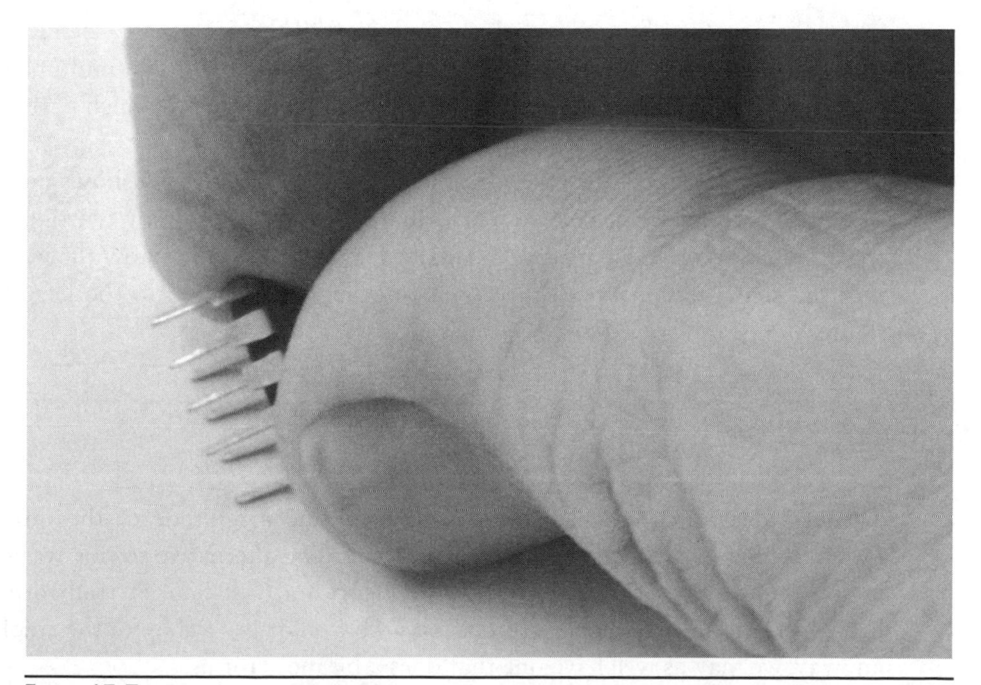

FIGURE 15-7 Bending the leads of the IC.

Figure 15-8 Protoshield with IC inserted.

and right signals or have three color-coded wires. If this is the case, then left and right usually will be white and red, respectively. If in doubt, use a multimeter in continuity buzzer mode to determine which lead is connected to which part of the audio plug.

Figure 15-9*a* shows the type of audio lead used. Figure 15-9*b* shows one-half of the lead stripped. The next step is to join the wires up again by twisting them together, as shown in Figure 15-9*c*. Finally, Figure 15-9*d* shows how the wires fit into the screw terminal with the common-ground connection to the left of the figure.

Software

The use of a hardware filter chip to separate the magnitude of the different frequency bands greatly simplifies the software. The alternative to this would be to use an amplifier chip and a software algorithm (fast Fourier transform) to produce the frequency analysis, but because we would need a chip for the amplifier anyway, we may as well have one that does a bit more for us.

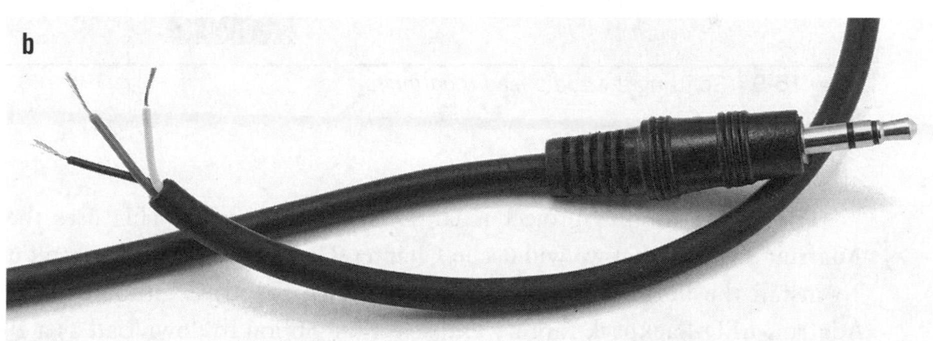

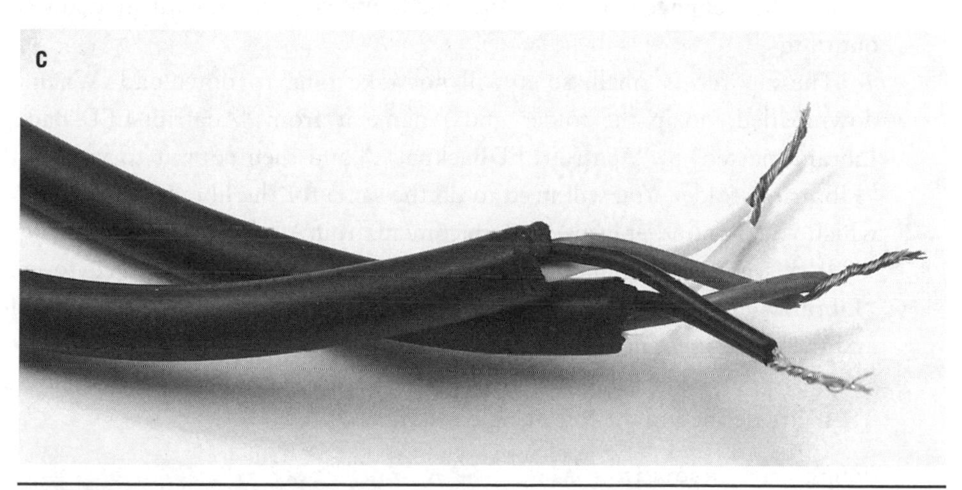

FIGURE 15-9 Splitting the audio lead.

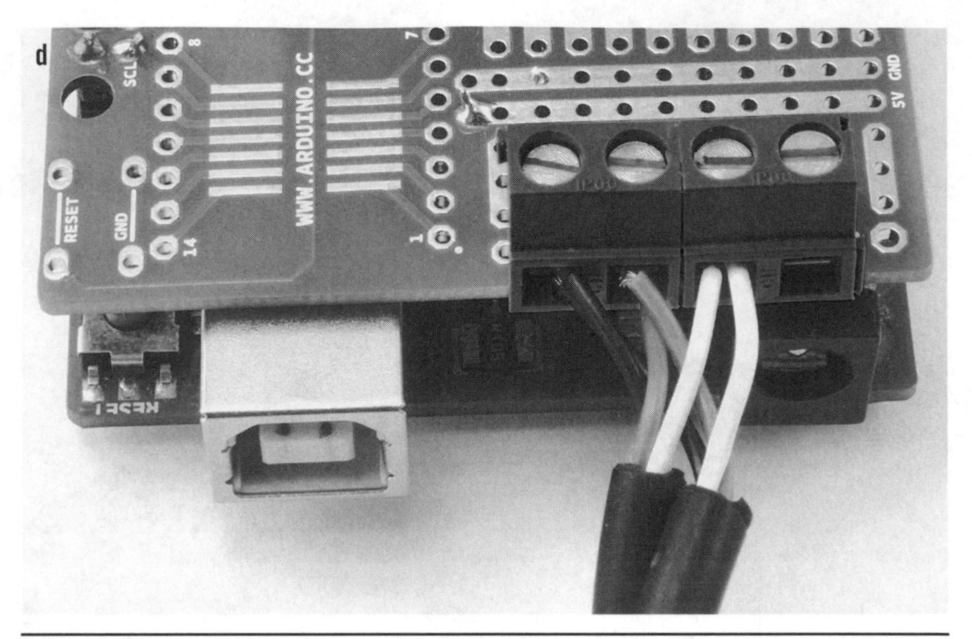

FIGURE 15-9 Splitting the audio lead (*continued*).

The sketch for this project is ch_15_spectrum_display. This uses the same Adafruit libraries that we will use in Chapter 21 to drive the LED matrix display. To install the libraries for the LED module, go to https://github.com/adafruit/Adafruit-LED-Backpack-Library and select the option to download as a zip. The icon on the webpage is labeled "Zip" and looks like a cloud with an arrow coming out of it.

The zip file is small, so it will not take long to download. When it has downloaded, unzip the folder and rename it from "Adafruit-LED-Backpack-Library-master" to "AdafruitLEDBackpack," and then copy it to your Arduino "Libraries" folder. You will need to do the same for the library "Adafruit_GFX," which you can find at https://github.com/adafruit/Adafruit-GFX-Library.

If this is the first time you have added a library, you will need to create a "Libraries" folder by opening the "Arduino" folder in your "Documents" folder and then creating a new folder called "Libraries."

Three pins are used in this project, two digital outputs and one analog input. These are defined in the following constants:

```
const int resetPin = A1;
const int strobePin = A2;
const int analogIn = A0;
```

The colors displayed at different magnitudes will vary, with green for low signals, yellow for medium, and red for high signals. These colors are specified in the following array:

```
const int colors[] = {LED_GREEN, LED_GREEN, LED_GREEN,
LED_YELLOW, LED_YELLOW, LED_YELLOW, LED_RED, LED_RED};
```

To create the display object, the following line of code is used:

```
Adafruit_BicolorMatrix matrix = Adafruit_BicolorMatrix();
```

The `setup` function is as follows. It simply defines the pin modes and initializes the display.

```
void setup()
{
  pinMode(resetPin, OUTPUT);
  pinMode(strobePin, OUTPUT);
  matrix.begin(0x70);
}
```

Most of the work for this project takes place in the following loop function:

```
void loop()
{
  pulsePin(resetPin, 100);
  matrix.clear();
  matrix.drawPixel(7, 0, colors[0]);
  for (int x = 6; x >= 0; x--)
  {
    pulsePin(strobePin, 50);
    int magnitude = analogRead(analogIn) >> 7;
    for (int y = 0; y <= magnitude; y++)
    {
      matrix.drawPixel(x, y, colors[y]);
    }
  }
  matrix.writeDisplay();
}
```

This function clears the display and then sends the reset signal to the MSGEQ7. It then puts a dot in the position (7, 0) of the display. Because the chip has only seven frequency bands but the display has eight, it looks better with something showing in the unused column.

There is then an outer `for` loop that will iterate over each of the frequency bands in turn, sending the `strobe` pulse to step onto the next channel. It then reads the value from the analog input pin.

An inner loop is used to effectively draw a line from the bottom of the display up the column that corresponds to the magnitude of the signal at that frequency band. The `colors` array is used to determine the color of the pixel.

The `loop` function uses the `pulsePin` command to generate a pulse on a pin supplied as its first argument for the duration in microseconds specified in its second.

```
void pulsePin(int pin, int duration)
{
  digitalWrite(pin, HIGH);
  delayMicroseconds(duration);
  digitalWrite(pin, LOW);
  delayMicroseconds(100);
}
```

Using the Project

You can test the project using a smartphone signal-generator app, as shown in Figure 15-1, by connecting just one of the leads to the phone. You could connect the other lead to your amplifier's line-in input, and you will hear the tone from the signal generator as well as being able to see it in the spectrum display.

Try varying the frequency of the tone, and you should see the column corresponding to the frequency of the signal generator show the strongest signal. When you are happy that everything is working as it should, then connect one end of the lead to your sound source (i.e., phone, MP3 player, or computer) and the other to your amplifier or powered speakers, and you should see the spectrum display jump around in time with the music.

Summary

This is the last of the audio projects. In the next section, you will find a selection of projects that all use the Internet in some way.

PART FOUR

Internet

CHAPTER 16

E-Mail Notifier

Difficulty: ★	Cost guide: $60

This wireless e-mail notifier combines a low-cost RF transmitter with an Arduino Ethernet shield to allow a button press on the remote control to trigger a predefined e-mail message to be sent (Figure 16-1).

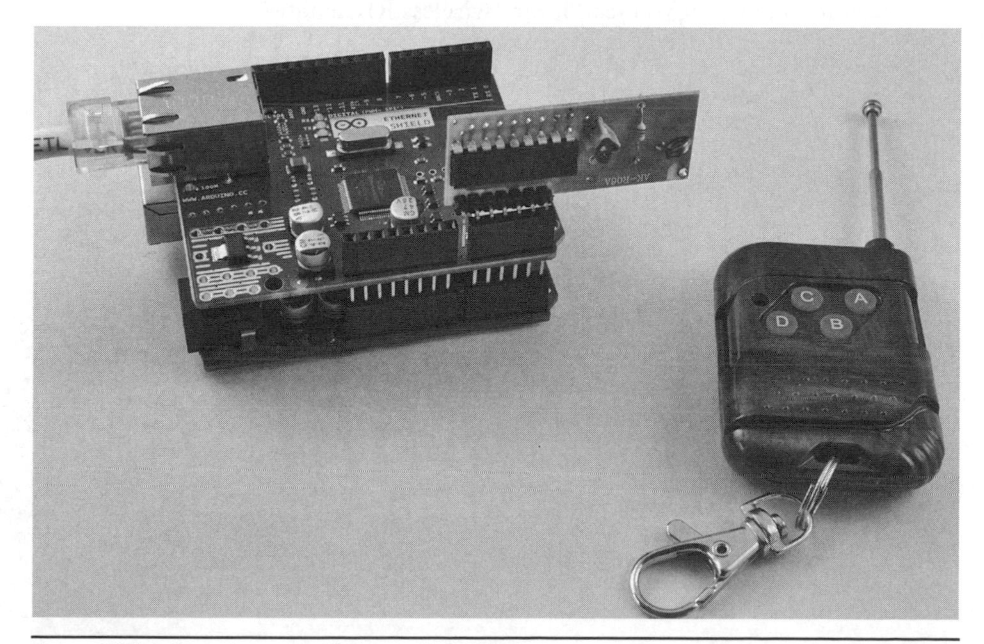

Figure 16-1 Remote-controlled e-mail notifier.

Parts

This project is compatible with both the Arduinos Uno and Leonardo. To build this project, you will need the following:

Part	Quantity	Description	Appendix
	1	Low-cost RF remote kit	eBay
	1	Arduino Ethernet shield[a]	M16
	1	12 V, 1 or 2 A power supply	M3

[a]This project will only work with Ethernet shields such as the official Arduino Ethernet shield or other variations that use the Wiznet chip set. Ethernet shields based on the ENC28J60 chip will not work. An alternative to using an Arduino and a separate Ethernet shield is to use a combination shield such as the Arduino Ethernet or the EtherTen from Freetronics.

The Ethernet shield uses 200 to 300 mA and may draw more current than the USB lead can comfortably cope with, so it is recommended that you use an external power supply. I found that I could get away without the external power supply when I used an Arduino Uno but not when I used a Leonardo. An external power supply will ensure better reliability.

Finding the same RF module will take a little searching on eBay. If you can find exactly the same type, it will fit neatly onto the analog header; otherwise, you will have to adapt the design a little. Figure 16-2 shows the module used. It should turn up on eBay if you search for “wireless RF remote.”

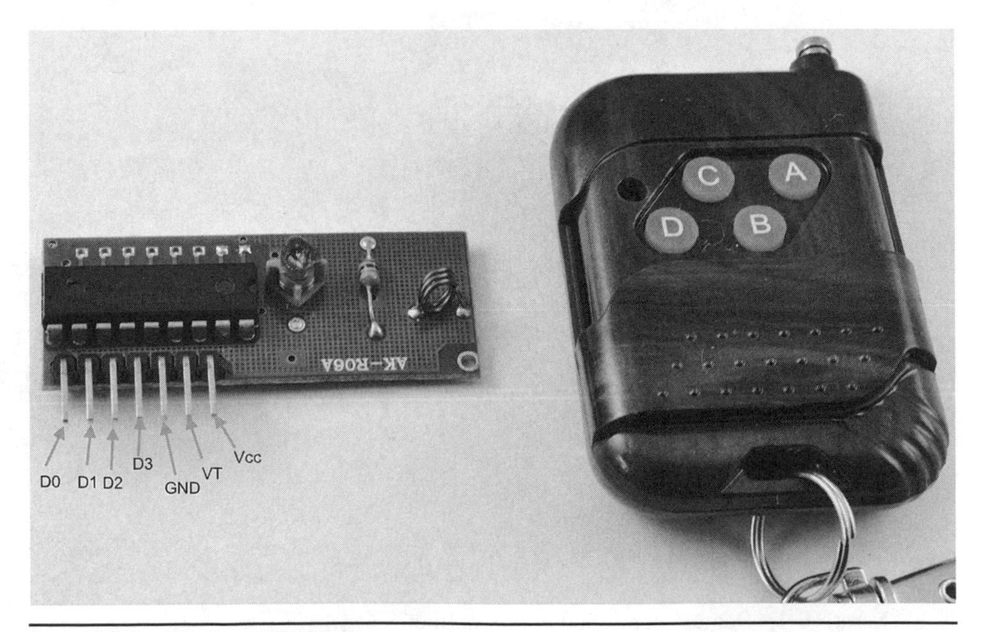

FIGURE 16-2 Wireless RF remote.

The wireless remote has four digital outputs that can be connected to digital inputs on the Arduino. Each of these receiver outputs is associated with one of the four buttons on the remote. When you press one of the buttons, the appropriate digital pin toggles between `HIGH` and `LOW`. The VT pin will normally be `LOW`, but if any of the buttons are pressed, it will go `HIGH` for the duration of the button press.

Construction

This project is one that requires no soldering at all. The RF module will simply fit onto the Arduino Ethernet shield on the analog pins, with one of its pins hanging over into the gap between the two sets of headers, as shown in Figure 16-3.

In this configuration, we do lose the use of one of the output pins, but in this project, we are just going to use the VT pin, which will tell us if any of the buttons are pressed.

Software

To send an e-mail, you need to have the Arduino communicate over the Internet with a Simple Mail Transfer Protocol (SMTP) server. This is notoriously difficult from an Arduino, which lacks the ability to use Secure Socket connections. Therefore, to simplify this process, we will cheat and use an SMPT service called

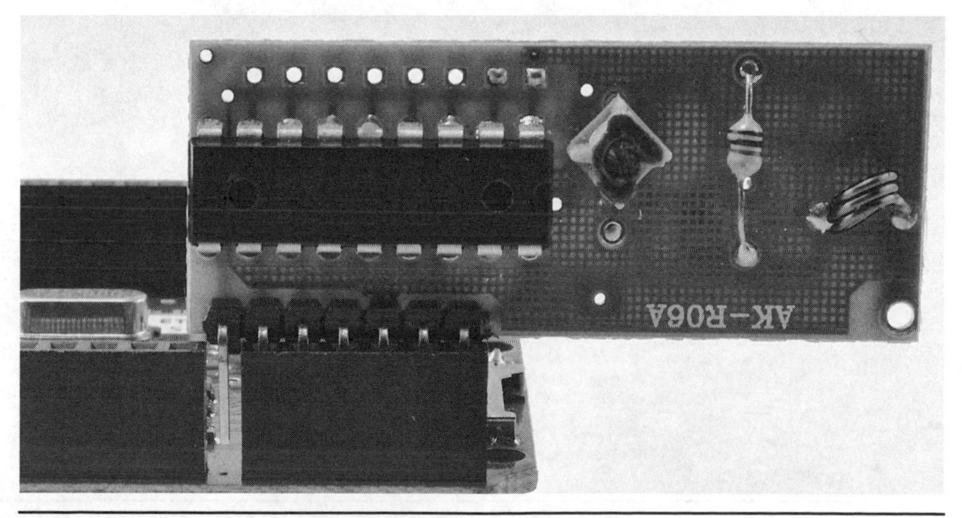

Figure 16-3 Attaching the RF receiver.

smtp2go. At the time of this writing, this service is free as long as you stay under the limit of 20 e-mails a day.

The first step is to create an account with smpt2go. Figure 16-4 shows the signup page. Be sure to select the free plan.

Having signed up, you will then need to set up an SMTP username and password. This is in addition to the username and password for the smtp2go website login, although you can enter the same details if you want. Figure 16-5 shows the webpage for adding this information.

Your Arduino is going to need to contain the SMTP username and password that you have just entered. This is not kept as plain text but rather is base64 encoded. A lot of people think that base64 makes your password secure because it is somehow encrypting it. This really is not the case. You can just paste a base64 string into an online converter, and it will give you back the plain-text password.

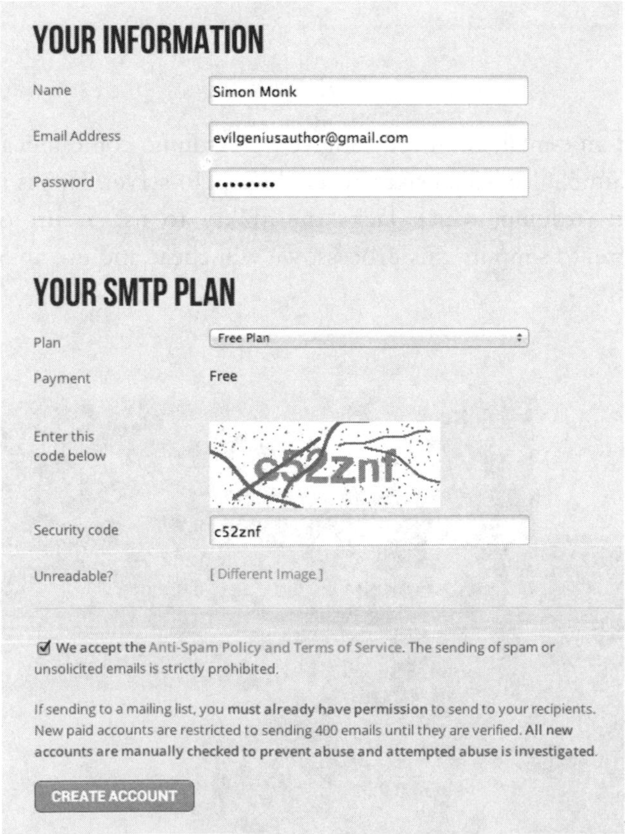

YOUR INFORMATION
Name: Simon Monk
Email Address: evilgeniusauthor@gmail.com
Password: ••••••••

YOUR SMTP PLAN
Plan: Free Plan
Payment: Free
Enter this code below
Security code: c52znf
Unreadable? [Different Image]

We accept the Anti-Spam Policy and Terms of Service. The sending of spam or unsolicited emails is strictly prohibited.

If sending to a mailing list, you must already have permission to send to your recipients. New paid accounts are restricted to sending 400 emails until they are verified. All new accounts are manually checked to prevent abuse and attempted abuse is investigated.

CREATE ACCOUNT

FIGURE 16-4 Signing-up with smtp2go.

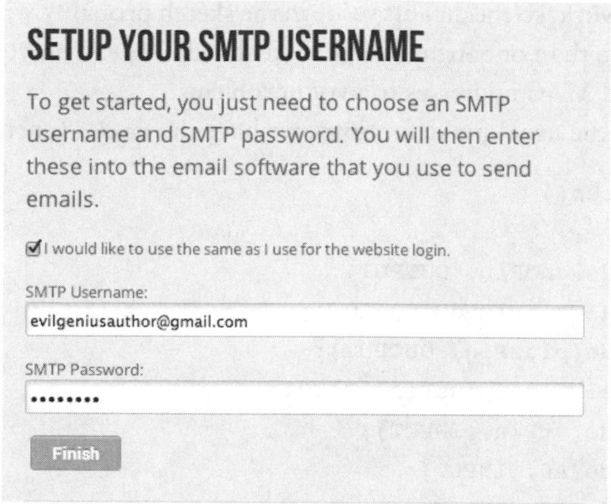

FIGURE 16-5 Adding an SMTP account.

However, the smtp2go service does require that the username and password be base64 encoded, so you will need to type each, in turn, into a base64 encoding tool. These are available online, for example, at http://base64-encoder-online.waraxe.us/.

You will now need to modify the first few lines of the sketch ch_16_email_notifier. In particular, you will need to modify the value of the following variables before you upload the sketch to your Arduino.

```
char fromEmail[] = "myemail@somemail.com";
char toEmail[] = "myemail@somemail.com";
char emailBase64[] = "hasdfgtff57hg56=";
char passwordBase64[] = "jgortejtgwyydf=";
char message[] = "A Button was pressed";
```

The variables fromEmail and toEmail are the e-mail address that will appear in the e-mail's from field and the recipient of any notifications when you press the button, respectively. You will need to paste the base64 strings for your e-mail address and smtp2go password that you encoded earlier into the appropriate variables. Finally, you probably will want to change the message that will be sent.

Skipping back to the start of the file, you will see this byte array defined:

```
byte mac[] = { 0xDE, 0xAD, 0xBE, 0xEF, 0xFE, 0xED };
```

On newer Ethernet shields, you will find a number such as this on the box that you should use here. The values in this array just have to be unique within your local network, so the default value in the sketch probably will work too, but if you have more than one Arduino Ethernet shield in use, then you will have to change one of the MAC addresses to avoid problems.

After the usual pin definitions, we have the setup function:

```
void setup()
{
  pinMode(gndPin, OUTPUT);
  digitalWrite(gndPin, LOW);
  pinMode(plusPin, OUTPUT);
  digitalWrite(plusPin, HIGH);
  pinMode(vtPin, INPUT);
  pinMode(A0, INPUT);
  pinMode(A1, INPUT);
  pinMode(A2, INPUT);
  Serial.begin(9600);
  while (!Serial) {};
  Serial.println("Ready");
}
```

This sets the appropriate pin modes, setting the pin plusPin HIGH because this will be used to provide power to the RF receiver. It then starts serial communications. The use of serial is not strictly necessary, and you may choose to remove it once the project is working. It is there simply to provide feedback at each stage of the e-mail sending process, making it easier to debug.

The main loop function that follows relies on a number of helper functions to do its job:

```
void loop()
{
  if (digitalRead(vtPin))
  {
    Serial.println("Email Sending Triggered");
    connectToNetwork();
    connectToMailServer();
    sendMessage(message);
    if (!client.connected())
    {
      Serial.println("disconnecting.");
      client.stop();
    }
```

```
    delay(3000);
  }
  displayClientResponse();
}
```

If the VT pin is `HIGH`, indicating a button press on the remote control, then the function `connectToNetwork` is called to establish a network connection. Next, the function `connectToMailServer` is called to begin the communication process with the SMTP server. SMTP is a text-based protocol, which is great for debugging, because it means that you can see the conversation between the Arduino and the SMPT server at smtp2go. After this, the `sendMessage` function is called. The `if` statement that follows this disconnects the client when all the communication is complete.

Let's now look at the helper functions that were used here, starting with `connectToNetwork`:

```
void connectToNetwork()
{
  if (Ethernet.begin(mac))
  {
    Serial.println("Connected with DHCP");
    for (byte thisByte = 0; thisByte < 4; thisByte++)
    {
      Serial.print(Ethernet.localIP()[thisByte], DEC);
      Serial.print(".");
    }
  }
  delay(3000);
}
```

Most of this function is concerned with displaying the Internet Protocol (IP) address of the Arduino when it connects to the network. Similarly, the `connectToMailServer` function is mostly concerned with displaying progress or failure in the serial monitor. The actual code that makes the connection is the call to `client.connect`, which returns `TRUE` if the connection is successful.

```
void connectToMailServer()
{
  Serial.println("connecting to SMPT Server...");
  if (client.connect(smtp, port))
  {
```

```
    Serial.println("Connected to mail server");
    delay(500);
  }
  else
  {
    Serial.println("Connection to mail server failed");
  }
}
```

The function `sendMessage` sends a sequence of commands to the SMTP server, first to authenticate and then to send the actual message:

```
void sendMessage(char message[])
{
    client.println("EHLO kltan");
    client.println("AUTH LOGIN");
    client.println(emailBase64);
    client.println(passwordBase64);
    client.print("MAIL FROM:");
    client.println(fromEmail);
    client.print("RCPT TO:");
    client.println(toEmail);
    client.println("DATA");
    client.print("from:");
    client.println(fromEmail);
    client.print("to:");
    client.println(toEmail);
    client.println("SUBJECT: Testing");
    client.println();
    client.println(message);
    client.println(".");
    client.println("QUIT");
}
```

The final function in the sketch is `displayClientResponse`. This is called every time around the loop. When there are responses from the server, these are echoed to the serial monitor.

```
void displayClientResponse()
{
```

```
    if (client.available())
    {
      char c = client.read();
      Serial.print(c);
    }
  }
```

Using the Project

Although in this case the e-mail sending is triggered by pressing a button on the remote, it could just as easily be triggered by a passive infrared (PIR) motion detector or any other kind of sensor. We have used a separate Arduino and Ethernet shield, but the project is a lot neater if you use a board such as the EtherTen from Freetronics, which is an Uno clone complete with Ethernet hardware. Figure 16-6 shows the project using an EtherTen.

The EtherTen also can be used with Power Over Ethernet (PoE), which would reduce the wiring considerably. Don't forget that the free smpt2go account allows only 20 e-mails a day.

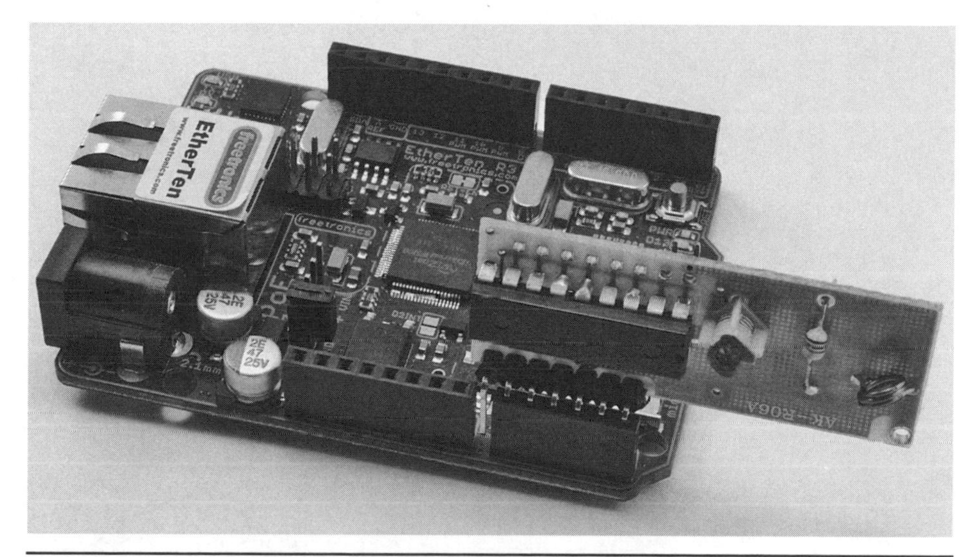

FIGURE 16-6 Using an all-in-one Ethernet board.

Summary

As a project, this could have been in the "Security" part of this book because it has applications for remote notifications of intruder alarm triggering or security for an elderly relative whose younger relatives may be alerted most easily by an e-mail.

CHAPTER 17

Weather Data Feed

Difficulty: ★	Cost guide: $50

This project uses a web service to retrieve weather information and display it on an LED matrix display (Figure 17-1).

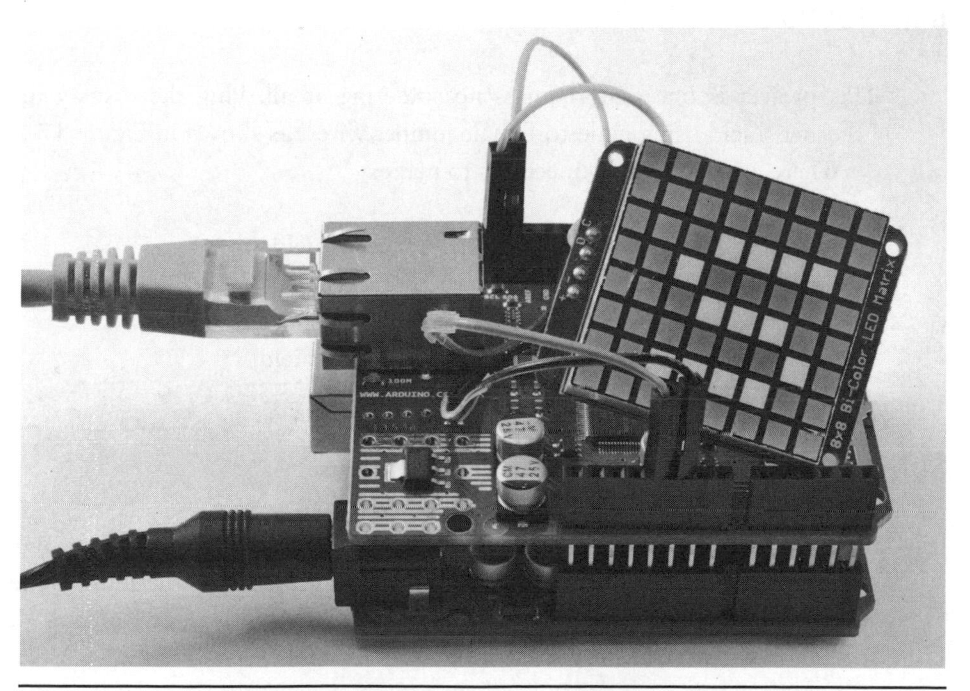

FIGURE 17-1 Displaying weather data.

Parts

This project is compatible with both the Arduinos Uno and Leonardo. To build this project, you will need the following:

Part	Quantity	Description	Appendix
	1	Arduino Ethernet shield[a]	M16
	1	Adafruit bicolor LED matrix	M4
	4	Male-to-female jumper wires	H6

[a]*This project will only work with Ethernet shields such as the official Arduino Ethernet shield or other variations that use the Wiznet chip set. Ethernet shields based on the ENC28J60 chip will not work. An alternative to using an Arduino and a separate Ethernet shield is to use a combination shield such as the Arduino Ethernet or the EtherTen from Freetronics.*

The Ethernet shield uses 200 to 300mA and may draw more current than the USB lead can comfortably cope with, so it is recommended that you use an external power supply. I found that I could get away without the external power supply when I used an Arduino Uno but not when I used a Leonardo. An external power supply will ensure better reliability.

Construction

This project is one that requires no soldering at all. Plug the display into the Ethernet shield using male-to-female jumper wires, as shown in Figure 17-2.

There are just four connections to make:

- + on the display to 5 V on the Ethernet shield
- – on the display to GND on the Ethernet shield
- D on the display to SDA on the Ethernet shield
- C on the display to SCL on the Ethernet shield

Software

The software uses the "Ethernet" library that is included in the Arduino IDE. The sketch also uses three libraries to drive the display. These are the same as the ones we used in Chapter 15, so please refer to the "Software" section of that chapter for details on installing the Adafruit display libraries.

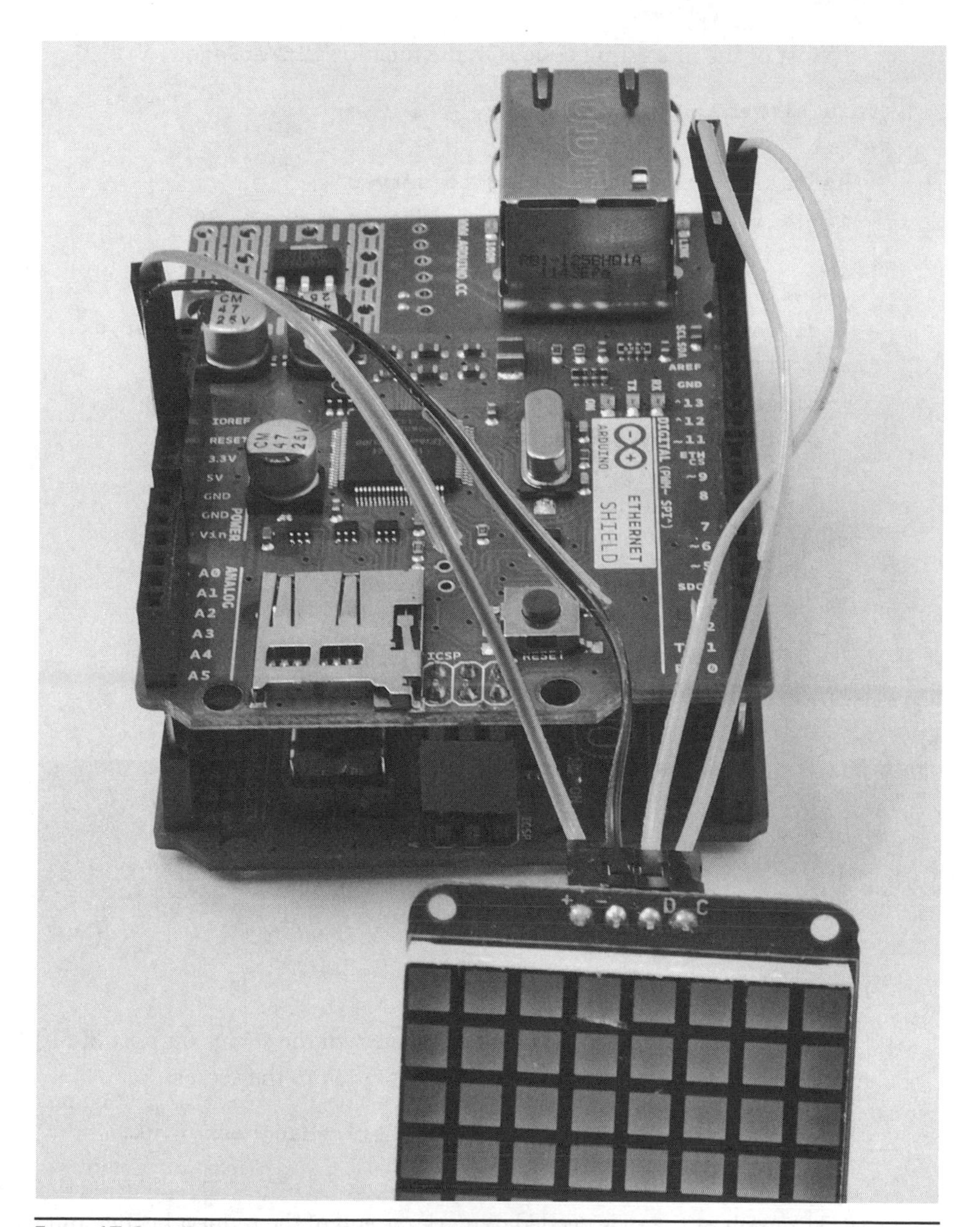

Figure 17-2 Wiring the display to the Ethernet shield.

The sketch (ch_17_weather_feed) has some similarities with that of Chapter 16, especially when it comes to establishing the network connection and sending the request to the server. However, in this case, the server is not an SMTP e-mail server but rather a web service that returns data about the weather for a particular location.

Most of the interesting code is in the function `hitWebPage`:

```
void hitWebPage()
{
  Serial.println("Connecting to server");
  if (client.connect("api.openweathermap.org", 80))
  {
    Serial.println("Connected to Server");
    client.println("GET /data/2.5/weather?q=Manchester,uk
      HTTP/1.0");
    client.println();
    while (client.connected())
    {
      if (client.available())
      {
        Serial.println("Reading Response");
        client.findUntil("description\":\"", "\0");
        String desc = client.readStringUntil('\"');
        if (desc.length() > 0)
        {
          desc.toCharArray(message, messageLength);
        }
      }
    }
    client.stop();
  }
}
```

The first step is to get the client to connect to the server on port 80. If this is successful, then the page request header is written to the server:

```
client.println("GET /data/2.5/weather?q=Manchester,uk
    HTTP/1.0");
```

The extra `println` is needed to mark the end of the request header and trigger a response from the server.

To wait for the connection, the `if` statement inside the `while` loop detects when data are available to be read. Reading the data stream directly avoids the need to capture all the data into memory. The data are in Java Script Object Notation (JSON) format:

```
{"coord":{"lon":-2.23743,"lat":53.480949},"sys":{"country":"GB",
"sunrise":1371094771,"sunset":1371155927},"weather":[{"id":520,"
main":"Rain","description":"light intensity shower rain","icon":
"09d"}],"base":"global stations","main":{"temp":284.87,"pressure
":1009,"humidity":87,"temp_min":283.15,"temp_max":285.93},"wind"
:{"speed":5.1,"deg":270},"rain":{"1h":0.83},"clouds":{"all":40},
"dt":1371135000,"id":2643123,"name":"Manchester","cod":200}
```


We are going to chop out the part of the text from description followed by a colon and then double quotes until the next double quote using the `findUntil` and `readStringUntil` functions. The `findUntil` function just ignores everything from the server until the matching string is found. The `readStringUntil` function then reads all the subsequent text until the double quote character.

We want the sketch to periodically display the data from the web service, but we will want to repeat the message, scrolling it across the display more frequently than we hit the web service. It would be irresponsible of us to be connecting to the web service every few seconds, so we use a variable `lastServerContact` to keep track of when we last went to the server and only go again when the value in the constant `serverCheckPeriod` has elapsed—in this case, after 60 seconds. Even this is a bit frequent, and you might like to change it to every 10 minutes once you know that everything is working.

```
void loop()
{
  long now = millis();
  if (now > lastServerContact + serverCheckPeriod)
  {
    hitWebPage();
    lastServerContact = now;
  }
  displayMessage();
}
```

Every time the web service is asked for its weather data, the salient part of the response is copied into the string array message so that the last response received from the weather service can be displayed by the `displayMessage` function that is called from the main `loop` function.

```
void displayMessage()
{
  matrix.setTextWrap(false);  // scroll
```

```
    matrix.setTextSize(1);
    matrix.setTextColor(LED_GREEN);
    for (int x = 7; x >= -messageLength*8; x--) {
      matrix.clear();
      matrix.setCursor(x, 0);
      matrix.print(message);
      matrix.writeDisplay();
      delay(100);
    }
}
```

The `displayMessage` function achieves the scrolling effect by repeatedly rewriting the message while changing the offset variable `x`.

Using the Project

The web request has a request parameter that is currently set as `/data/2.5/weather?q=Manchester,uk`. This is set to a location near my home. I guess I could just look out the window, but that is much too easy. You can change location to, for example, `/data/2.5/weather?q=San%20Francisco,us` to show the weather in San Francisco. Note the use of `%20` to separate the words of the city name.

Summary

This is a fun little project and one that could easily be adapted to other kinds of web services as well as other types of displays. In the final two chapters of this section, we will use the Arduino as a web server rather than a client.

CHAPTER 18

Network-Controlled Switch

Difficulty: ★	Cost guide: $70

Arduino is used a lot in home automation systems, usually with an Ethernet or WiFi card. The mainstay of home automation is to be able to turn things on and off remotely. Having a networked Arduino allows you all sorts of possibilities in this respect because if the Arduino is acting as a web server, then you can interact with it using a browser on any device in the network. Therefore, suddenly, you can use a phone or a tablet to turn things on and off.

This project uses a networked Arduino attached to a device called the *PowerSwitch Tail* that makes it really easy and safe to switch alternating-current (ac) appliances on and off. Figure 18-1 shows the Arduino with Ethernet shield connected to the PowerSwitch Tail, and Figure 18-2 shows the web interface for this project in the web browser of an Android phone.

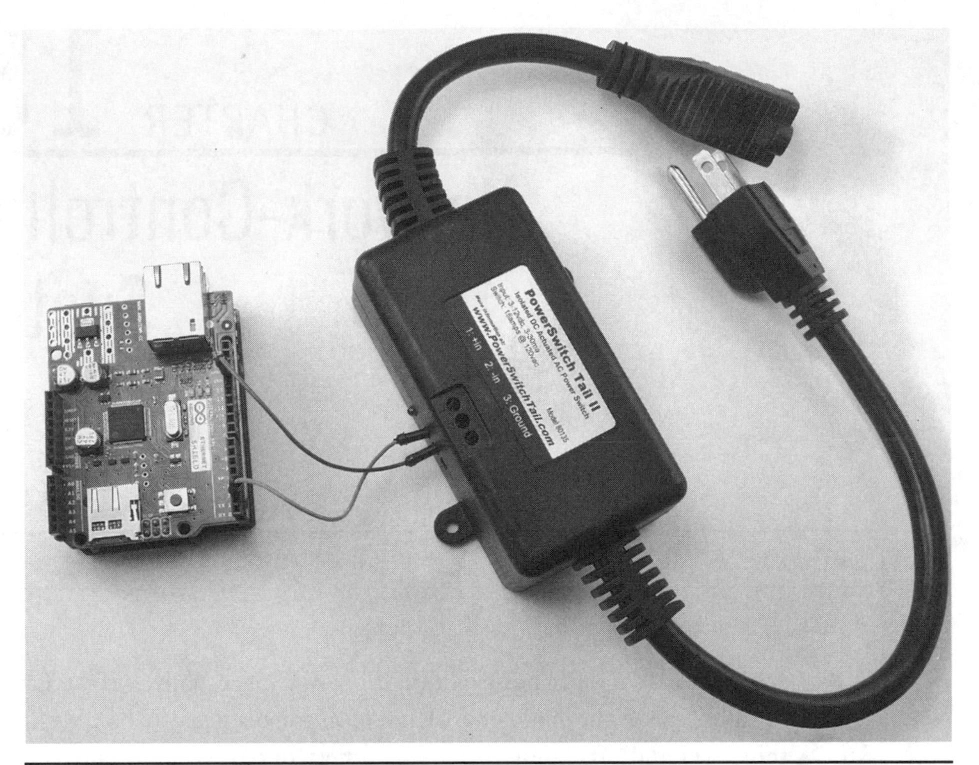

Figure 18-1 Network-controlled switch.

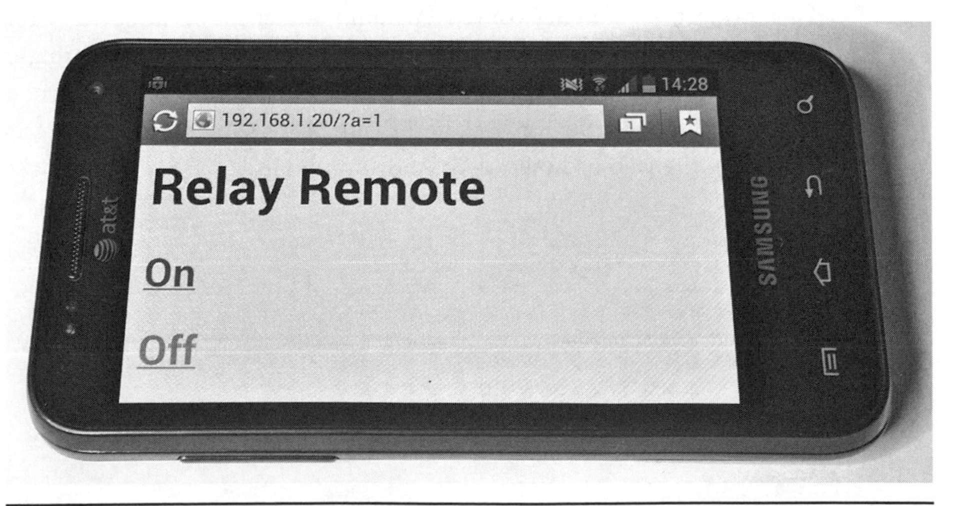

Figure 18-2 Web interface to the switch.

Parts

This project is compatible with both the Arduinos Uno and Leonardo. To build this project, you will need the following:

Part	Quantity	Description	Appendix
	1	Arduino Ethernet shield[a]	M16
	1	PowerSwitch Tail	M17
	2	Male-to-male jumper wires	H11

[a]This project will only work with Ethernet shields such as the official Arduino Ethernet shield or other variations that use the Wiznet chip set. Ethernet shields based on the ENC28J60 chip will not work. An alternative to using an Arduino and a separate Ethernet shield is to use a combination shield such as the Arduino Ethernet or the EtherTen from Freetronics.

The Ethernet shield uses 200 to 300 mA and may draw more current than the USB lead can comfortably cope with, so it is recommended that you use an external power supply. I found that I could get away without the external power supply when I used an Arduino Uno but not when I used a Leonardo. An external power supply will ensure better reliability.

Construction

This is another project that requires no soldering at all. Use the jumper wires to connect one of the GND connections of the Arduino to the screw terminal marked with a minus sign ("–") on the PowerSwitch tail and the Arduino D2 connection into the screw terminal marked with a plus sign ("+") on the PowerSwitch Tail.

Software

The only library that this project uses is the "Ethernet" library included in the Arduino IDE.

Let's break this sketch down and look at it a section at a time. The first part of this sketch (ch_18_network_switch) is much the same as the sketches in Chapters 16 and 17, using the same `connectToNetwork` function. Please refer to the descriptions of the software for these two chapters if you need more information.

The loop function is responsible for servicing any requests that come to the web server from a browser. If a request is waiting for a response, then calling server.available will return us a client. If client exists (tested by the first if statement), then we can determine whether it is connected to the web server by calling client.connected.

```
void loop()
{
  EthernetClient client = server.available();
  if (client)
  {
    while (client.connected())
    {
      readHeader(client);
      if (! pageNameIs("/"))
      {
        client.stop();
        return;
      }
      digitalWrite(relayPin, valueOfParam('a'));
      client.println("HTTP/1.1 200 OK");
      client.println("Content-Type: text/html");
      client.println();

      // send the body
      client.println("<html><body>");
      client.println("<h1>Relay Remote</h1>");

      client.println("<h2><a href='?a=1'/>On</a></h2>");
      client.println("<h2><a href='?a=0'/>Off</a></h2>");
      client.println("</body></html>");

      client.stop();
    }
  }
}
```

We will come to the readHeader function later. This function and pageNameIs are used to determine that the browser is actually contacting the page for setting the relay. This is so because browsers will often send two requests to a server page, one to try to find an icon for the website and a second to the page itself. This code allows us to ignore the icon request.

The next line sets the relay pin using `digitalWrite`. The value it sets the output to is whatever value the request parameter `a` is set to. This will be either `1` or `0`. The next three lines of code print out a return header. This just tells the browser what type of content to display—in this case, just HyperText Markup Language (HTML).

Once the header has been written, it just remains to write the remaining HTML back to the browser. This must include the usual `<html>` and `<body>` tags and also includes a `<h1>` header tag and two `<h2>` tags that are also hyperlinks to this same page but with the request parameter `a` set to either `0` or `1`.

Finally, `client.stop` tells the browser that the message is complete, and the browser will display the page.

```
void readHeader(EthernetClient client)
{
  // read first line of header
  char ch;
  int i = 0;
  while (ch != '\n')
  {
    if (client.available())
    {
      ch = client.read();
      line1[i] = ch;
      i ++;
    }
  }
  line1[i] = '\0';
  Serial.println(line1);
}
```

The final three functions in the sketch are general-purpose functions that I tend to use over and over again when making an Arduino web server such as this. The first, `readHeader`, reads the header of the response coming from the browser into the buffer `line`. We can then use this in the next two functions:

```
boolean pageNameIs(char* name)
{
   // page name starts at char pos 4
   // ends with space
   int i = 4;
   char ch = line1[i];
```

```
    while (ch != ' ' && ch != '\n' && ch != '?')
    {
      if (name[i-4] != line1[i])
      {
        return false;
      }
      i++;
      ch = line1[i];
    }
    return true;
}
```

The function `pageNameIs` returns `TRUE` if the page name part of the header matches the argument supplied. This is what we use in the `loop` function to ignore the icon request from the browser.

```
int valueOfParam(char param)
{
  for (int i = 0; i < strlen(line1); i++)
  {
    if (line1[i] == param && line1[i+1] == '=')
    {
      return (line1[i+2] - '0');
    }
  }
  return 0;
}
```

The `valueOfParam` allows us to read the value of the request parameter supplied as an argument. This is much more restricted than the kind of request parameter you will be used to if you have done any web programming. First, the request parameter name must be a single character, and second, its value must be a single digit between `0` and `9`. The function will return the value or `0` if there is no parameter of that name.

Using the Project

This is one of those projects that can be adapted for all sorts of purposes. Open up a browser on your computer, tablet, or smartphone, and navigate to the IP

Figure 18-3 IP address of the Arduino in the serial monitor.

address. To see the IP address allocated to the Arduino by your network, open the serial monitor, and it will be displayed after the Arduino connects (Figure 18-3).

For my router, this IP address is `http://192.168.1.20`. You should then see the web page of Figure 1-2 displayed. Click on the "On" button, and the LED on the PowerSwitch tail will turn on. The page will then reload in the browser, and clicking "Off" will turn the switch off again. Plug one end of the PowerSwitch tail into an ac outlet and an appliance such as a desk lamp into the other end, and you will be able to turn the lamp on and off from the browser on your phone, tablet, or PC.

Hint: Fixing the IP Address of the Arduino

If this project stops working inexplicably, then it may be that the Arduino has been assigned a different IP address. There is a simple way to keep the same IP address reserved for the Arduino. You need to change the sketch so that instead of using the Dynamic Host Configuration Protocol (DHCP) to allow the network to allocate the IP address for the Arduino, you fix the IP address (make sure that you pick one that's not in use) by adding the following line of code after the `mac` array declaration:

```
byte ip[] = { 192, 168, 1, 20 };
```

Substitute your own IP address into this array, but an address allocated for the Arduino by DHCP will be a good bet. You also will need to replace the function `connectToNetwork` with the following version:

```
void connectToNetwork()
{
  Ethernet.begin(mac, ip);
  delay(3000);
}
```

The modified version of the sketch is available as ch_18_network_switch_fixed_ip.

Summary

You could, of course, attach a whole load of PowerSwitch tails to your Arduino to control more than one device. If you do this, you will need to modify your sketch accordingly. In Chapter 19, we will look at using an Arduino web server to serve up sensor readings.

CHAPTER 19

Network Temperature and Humidity Sensor

Difficulty: ★	Cost guide: $50

This project uses a low-cost combined temperature and humidity sensor module (DHT11) and an Ethernet shield to make a web interface that displays the temperature in both degrees Centigrade and Fahrenheit and the relative humidity. The page refreshes every second. Figure 19-1 shows the project hardware, and Figure 19-2 shows the web interface.

Parts

This project is compatible with both the Arduinos Uno and Leonardo. To build this project, you will need the following:

Part	Quantity	Description	Appendix
	1	Arduino Ethernet shield[a]	M16
	1	DHT11 temperature and humidity sensor	M18

[a]This project will only work with Ethernet shields such as the official Arduino Ethernet shield or other variations that use the Wiznet chip set. Ethernet shields based on the ENC28J60 chip will not work. An alternative to using an Arduino and a separate Ethernet shield is to use a combination shield such as the Arduino Ethernet or the EtherTen from Freetronics.

The Ethernet shield uses 200 to 300 mA and may draw more current than the USB lead can comfortably cope with, so it is recommended that you use an external power supply. I found that I could get away without the external power supply

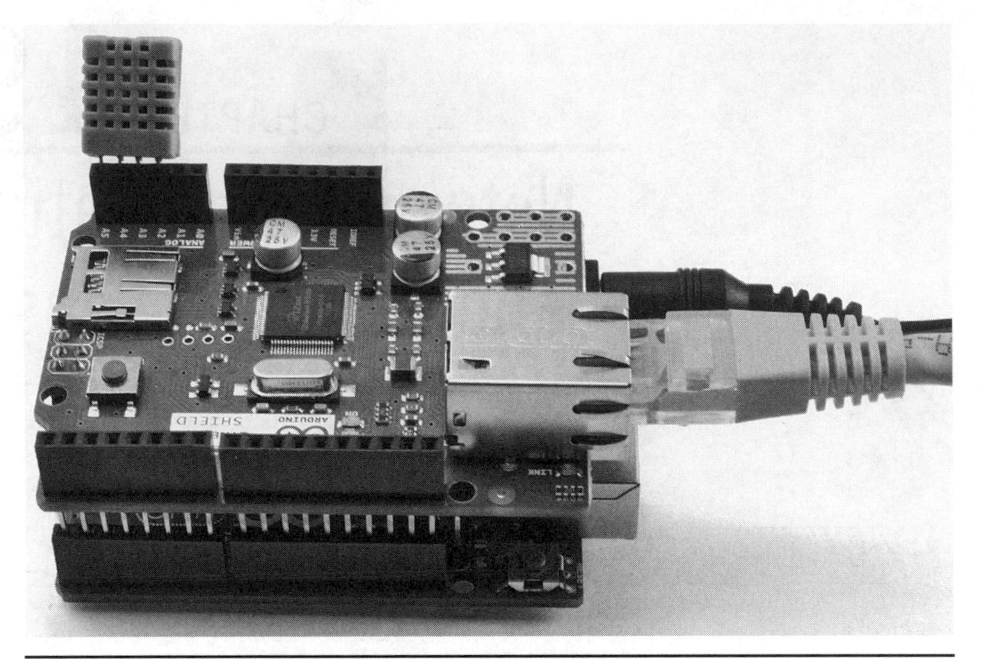

Figure 19-1 Network temperature and humidity sensor.

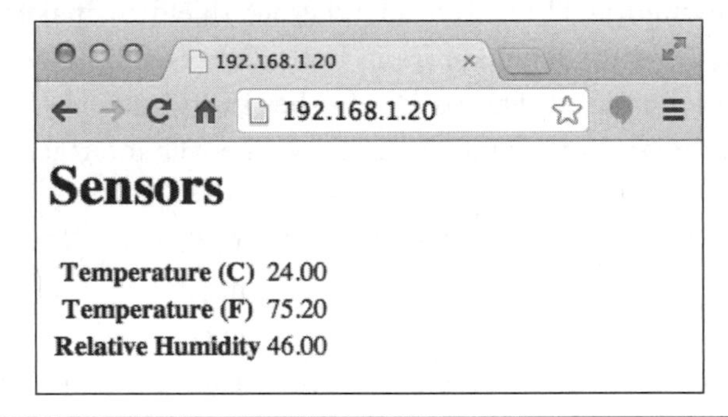

Figure 19-2 Web interface to the network sensor.

when I used an Arduino Uno but not when I used a Leonardo. An external power supply will ensure better reliability.

Construction

The DHT11 requires a pull-up resistor between its output and 5 V. It really didn't seem worth having a shield to accommodate one resistor and the DHT11 module, so the design uses two of the analog pins as outputs to supply 5 V to the module.

Figure 19-3 Pinout of the DHT11.

To avoid the need for the pull-up resistor, the sketch turns on the internal pull-up resistor on the pin used as an input. Figure 19-3 shows the pinout of the DHT11. Note that one of the pins is not used at all.

This means that the module pins can just push into the right-hand four connectors of the analog section of the Arduino, as shown in Figure 19-1. If you find that the pins are not making good contact, put a little kink in them with pliers. Make sure that you get the module the right way around, and do not plug it in until after you have uploaded the sketch.

Software

There are a number of Arduino libraries for the DHT11. They are all slightly different, but the easiest to use and best thought out can be found at https://github.com/markruys/arduino-DHT. You will need to download the zip file (button on the right of the "Github" page) and then rename the folder to just "DHT" before saving it in your Arduino "Libraries" folder. Remember that you need to restart the Arduino IDE for new libraries to be recognized.

The sketch is ch_19_network_sensors and is available with all the book downloads. Much of the sketch is the same as the web server example from

Chapter 18. In fact, this sketch is simpler because it does not need to access the request parameters, but it does have a bit more HTML to serve back to the browser.

The `setup` function is listed below. This sets the Arduino pins on the analog connector to supply power to the module and also starts the network and server connections as well as initializing the DHT library.

```
void setup()
{
  pinMode(ignorePin, INPUT);
  pinMode(gndPin, OUTPUT);
  pinMode(plusPin, OUTPUT);
  digitalWrite(plusPin, HIGH);
  pinMode(dataPin, INPUT_PULLUP);
  Serial.begin(9600);
  while (!Serial){};
  connectToNetwork();
  server.begin();
  dht.setup(dataPin);
}
```

The `loop` function is quite long. The overall structure is the same as in Chapter 18, but now, as part of the process of generating the HTML to send back to the browser, we need to get some readings from the module. These are the lines of code that do this:

```
float tempC = dht.getTemperature();
float tempF = tempC * 9.0 / 5.0 + 32.0;
float humidity = dht.getHumidity();
```

The HTML that is generated also needs to include some Javascript to be run on whatever browser is displaying the readings so that the readings are automatically updated. The code that streams this HTML back to the browser is

```
    client.println("HTTP/1.1 200 OK");
    client.println("Content-Type: text/html");
    client.println();

    // send the body
    client.println("<html>");
    client.println("<head><script>setTimeout((function()
       {window.location.reload(true);}),
```

```
        1000)</script></head>");
      client.println("<body>");
      client.println("<h1>Sensors</h1>");

      client.println("<table>");
      client.println("<tr><th>Temperature (C)</th>");
      client.print("<td>"); client.print(tempC);
        client.println("</td>");
      client.println("</tr>");

      client.println("<tr><th>Temperature (F)</th>");
      client.print("<td>"); client.print(tempF);
        client.println("</td>");
      client.println("</tr>");

      client.println("<tr><th>Relative Humidity</th>");
      client.print("<td>"); client.print(humidity);
        client.println("</td>");
      client.println("</tr>");

      client.println("</table></body></html>");

      client.stop();
```

When writing code like this, it is very easy to miss a closing bracket, so a useful technique is to view the source of the served page in the browser. In this case, it will look something like this

```
<html>
<head><script>setTimeout((function(){window.location.
  reload(true);}), 1000)</script></head>
<body>
<h1>Sensors</h1>
<table>
<tr><th>Temperature (C)</th>
<td>24.00</td>
</tr>
<tr><th>Temperature (F)</th>
<td>75.20</td>
</tr>
```

```
<tr><th>Relative Humidity</th>
<td>54.00</td>
</tr>
</table></body></html>
```

Using the Project

You could, if you prefer, run longer leads to the module and have it away from the Arduino. If you do this, you probably will need to use a 4.7-kΩ resistor. This just needs to be connected from the data pin of the DHT11 to the Vcc (5 V) pin of the DHT11.

You probably have already discovered that the DHT11 is not terribly accurate. Higher-resolution versions of this module are available, and the library will work with these modules too. The DHT22 and RHT03 are devices to look for if you want better accuracy.

Summary

This is a project that could be adapted for other types of sensors. It also could be combined with the project of Chapter 18 to provide a server that both reports measurements and allows things to be switched on and off, all from a web interface. In Chapter 20, we will look at making a meter to measure the "ping" performance of your Internet connection.

CHAPTER 20

Pingometer

Difficulty: ★	Cost guide: $50

This project uses an analog voltmeter to display the ping time to a specified server (Figure 20-1).

Serious gamers often want to know what the ping value is of the game server to which they are connecting. The ping value gives an indication of how laggy the connection is going to be. The ping value is a number in milliseconds. Thus, a responsive connection to a server will have a low ping value (perhaps 50 milliseconds), and a slow connection could have a ping value of 500 milliseconds.

Parts

This project is compatible with both the Arduinos Uno and Leonardo. To build this project, you will need the following:

Part	Quantity	Description	Appendix
	1	Arduino Ethernet shield[a]	M16
	1	5 V panel meter	M19

[a]This project will only work with Ethernet shields such as the official Arduino Ethernet shield or other variations that use the Wiznet chip set. Ethernet shields based on the ENC28J60 chip will not work. An alternative to using an Arduino and a separate Ethernet shield is to use a combination shield such as the Arduino Ethernet or the EtherTen from Freetronics.

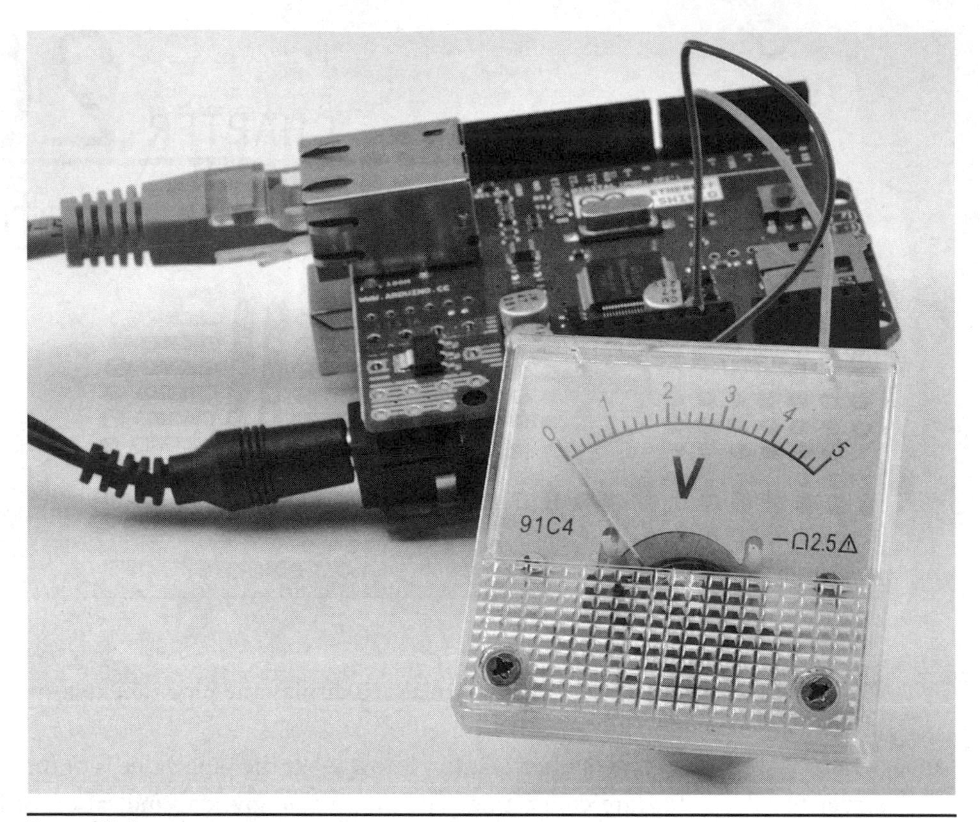

FIGURE 20-1 Pingometer.

The Ethernet shield uses 200 to 300 mA and may draw more current than the USB lead can comfortably cope with, so it is recommended that you use an external power supply. I found that I could get away without the external power supply when I used an Arduino Uno but not when I used a Leonardo. An external power supply will ensure better reliability.

Any kind of 5-V panel meter can be used in this project. Even a really large meter will use very little current and can be connected directly to an Arduino PWM output.

Construction

Construction is very easy for this project. In fact, you can get away without any soldering at all if you want to.

Step 1: Attach the Leads to the Meter

The panel meter is likely to have screw terminals (Figure 20-2).

To connect the meter to the Arduino, you need to attach some wires to the terminals. Solid-core wire is easiest for this, and it helps to make little loops on the ends that attach to the meter by bending them around the terminal posts and then lightly soldering the loop where the wire meets itself again. Put the loops over the terminal posts, and then do up the nuts with pliers.

Step 2: Connect the Meter to the Ethernet Shield

When you look at the connections on the back of the meter, one of the connections will be marked as positive and the other as negative. The negative connection needs connecting to one of the GND sockets of the Arduino, and the positive connection to the meter should be connected to D6.

If the solid-core wires feel a bit loose in the shield sockets, then put a little kink in the wires so that they make good contact.

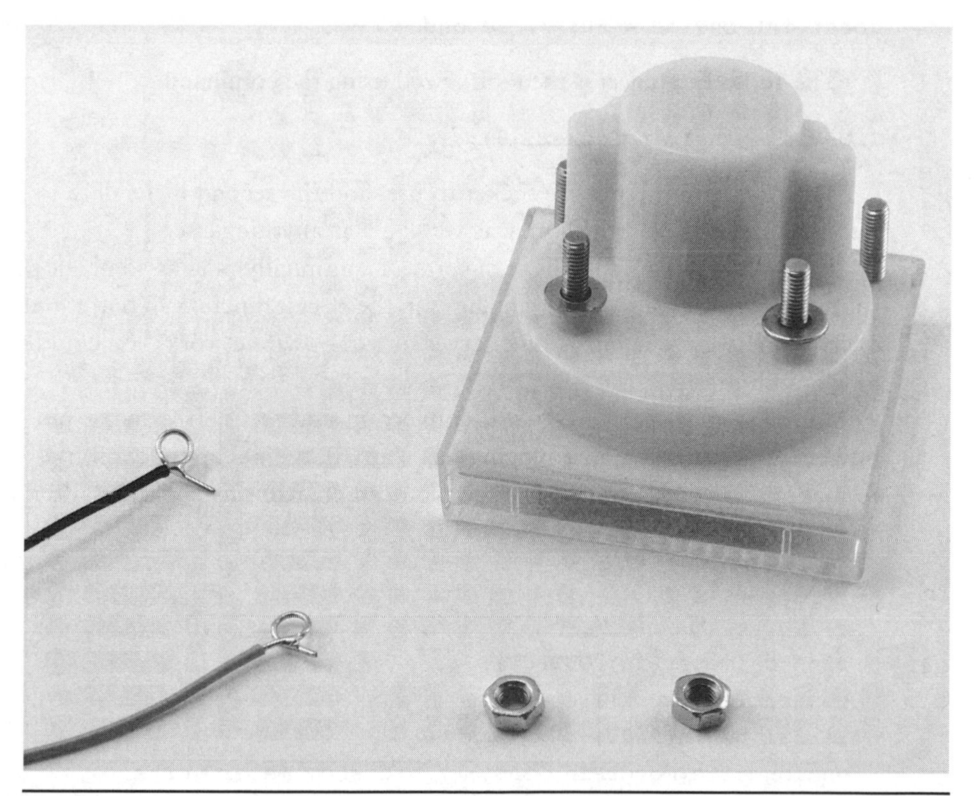

FIGURE 20-2 Connecting the wires to the meter.

Software

Fortunately for us, Blake Foster has written an excellent Arduino library for ping. You can find out more about it at http://playground.arduino.cc/Code/ICMPPing. This library does pretty much all the work for us apart from controlling the meter. The sketch for this project is ch_20_pingometer.

If you have worked through the previous chapters in this section, then you will be fairly familiar with Ethernet sketches. After the usual imports and definition of the MAC address array, the IP address is defined (`pingAddr`). This is the IP address of the server whose ping we want to test. This is defaulted to one of Google's IP addresses. In the next section, you will see how this can be altered.

```
IPAddress pingAddr(173, 194, 34, 103);
```

Two constants are then defined that control the scale of the meter and the length of time the Arduino will keep requesting pings from the server.

```
const int fullScalePing = 500;
const int onTime = 60; // seconds
```

The `ICMPPing` object is then initialized using the command

```
ICMPPing ping(pingSocket, 0);
```

The first parameter is the socket to use, and the second is the data to be put into the ping packets, which may as well be `0` as anything else.

It would be antisocial to have the project continually pinging someone's server, so the code to actually do the pinging is in the `setup` function so that it only runs once. If you were pinging a game server on your own network, you could simply ping continuously.

After establishing the network connection and setting the meter pin to full power (to demonstrate that pinging has started), the `setup` function uses a loop to make 60 seconds' worth of pings. You can change this period by altering the `onTime` constant.

```
void setup()
{
  pinMode(meterPin, OUTPUT);
  connectToNetwork();
  Serial.begin(9600);
  analogWrite(meterPin, 255); // to show meter works
```

```
  for (int i = 0; i < onTime * 2; i++)
  {
    ICMPEchoReply echoReply = ping(pingAddr, 4);
    if (echoReply.status == SUCCESS)
    {
      showPingReading(millis() - echoReply.data.time);
    }
    else
    {
      Serial.println("Ping Failed");
      analogWrite(meterPin, 255);
    }
    delay(500);
  }
  analogWrite(meterPin, 0);
}
```

Each time around the `for` loop, a ping is made, and if it is successful, the function `showPingReading` is called with the difference between the current time in milliseconds and the time that the ping returned.

The `showPingReading` function first of all prints the raw time to the serial monitor (for debugging purposes) and then calculates a scaled value between `0` and `255` to display the ping. The sensitivity of this can be adjusted by altering the value of the constant `fullScalePing`. With this variable set to its default value of `500`, a ping of 500 milliseconds will cause the needle to be hard over to the right. If your meter is calibrated in volts, then this will display 5. Thus, the scale would, in effect, read the ping time in tenths of a second.

```
void showPingReading(long ms)
{
  Serial.println(ms);
  long dutyCycle = ms * 255 / fullScalePing;
  if (dutyCycle < 0) dutyCycle = 0;
  analogWrite(meterPin, dutyCycle);
}
```

The `connectToNetwork` is identical to that used in the earlier projects in this section and uses the Dynamic Host Configuration Protocol (DHCP) to obtain an IP address for the Arduino.

Using the Project

This project will ping for 1 minute after it has been uploaded or powered up. To set it off pinging again, just press the "Reset" button on the Ethernet shield or the Arduino. The default code monitors Google. If you are a gamer, you are likely to want to change the IP address to that of the game server you are using. If you do not know the IP address of the server and only know it by its domain name, there are various ways that you can find the IP address. One way is to use a ping from your computer.

If you are using Windows, open a command prompt. On Linux or Mac, open a terminal session. Type the command `ping`, followed by the domain in which you are interested without "http://" on the front. You will see a response something like Figure 20-3. Here you can see that as part of the pinging process, the IP address of Google has been looked up.

```
Si — bash — 62×17
Simons-Mac-Pro:~ Si$ ping google.com
PING google.com (173.194.41.100): 56 data bytes
64 bytes from 173.194.41.100: icmp_seq=0 ttl=56 time=27.312 ms
64 bytes from 173.194.41.100: icmp_seq=1 ttl=56 time=26.769 ms
64 bytes from 173.194.41.100: icmp_seq=2 ttl=56 time=27.109 ms
64 bytes from 173.194.41.100: icmp_seq=3 ttl=56 time=26.752 ms
64 bytes from 173.194.41.100: icmp_seq=4 ttl=56 time=27.113 ms
64 bytes from 173.194.41.100: icmp_seq=5 ttl=56 time=27.183 ms
64 bytes from 173.194.41.100: icmp_seq=6 ttl=56 time=27.042 ms
64 bytes from 173.194.41.100: icmp_seq=7 ttl=56 time=27.428 ms
64 bytes from 173.194.41.100: icmp_seq=8 ttl=56 time=27.068 ms
64 bytes from 173.194.41.100: icmp_seq=9 ttl=56 time=27.160 ms
^C
--- google.com ping statistics ---
10 packets transmitted, 10 packets received, 0.0% packet loss
round-trip min/avg/max/stddev = 26.752/27.094/27.428/0.199 ms
Simons-Mac-Pro:~ Si$
```

Figure 20-3 Ping from Windows.

Summary

This is the last of the Internet-related projects. In Chapter 21, we will start a new section that looks at clocks of various types.

PART FIVE

Clocks

CHAPTER 21

LED Matrix Clock

Difficulty: ★★	Cost guide: $30

This project uses an LED matrix module and a real-time clock to display the time and date in scrolling text (Figure 21-1).

When the button is pressed, the clock toggles between displaying the time and displaying the date. The time is automatically set from your computer's clock when uploading the sketch if the button is pressed during the sketch upload.

Parts

To build this project, you will need the following:

Name	Quantity	Description	Appendix
	1	Protoshield printed circuit board (PCB) and header pins	A3, H1
	1	Adafruit bicolor LED module	M4
	1	Adafruit real-time clock (RTC) breakout	M5
		Solid-core wire	
S1	1	Click switch	C2
	2	Four-way header sockets[a]	H2

[a]The header socket is optional; you can solder the modules directly to the Protoshield if you prefer.

Figure 21-1 LED matrix clock.

Protoshield Layout

Figure 21-2 shows the schematic diagram for the project, and Figure 21-3 shows the Protoshield layout.

This project uses two modules from Adafruit. The real-time clock (RTC) module has a small battery that allows the clock to retain the time even when the power is disconnected. The LED module has an array of 8 × 8 LEDs, each of which can be red or green, or orange if both are on together. Both modules have a type of serial interface called I2C that makes them really easy to interface with the Arduino.

Construction

This is a pretty straightforward project to build. As always, start by attaching header pins to the Protoshield.

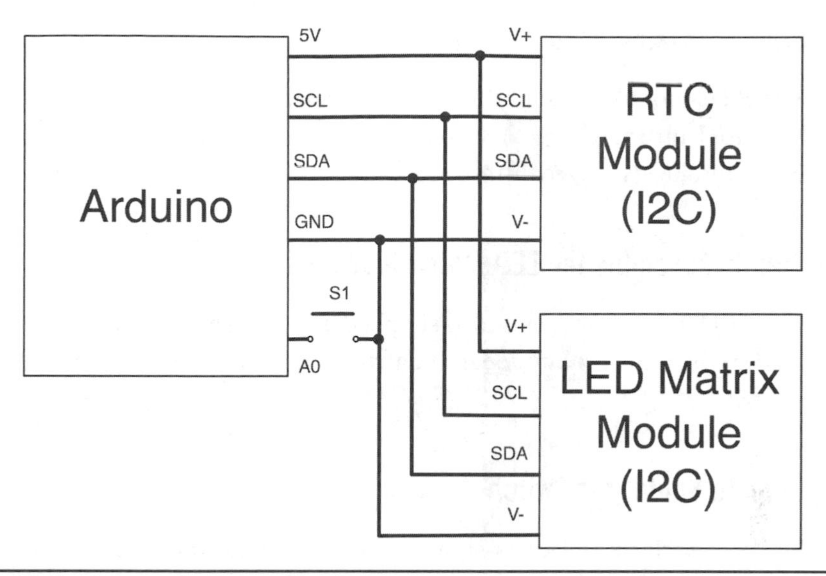

FIGURE 21-2 Schematic diagram for the clock.

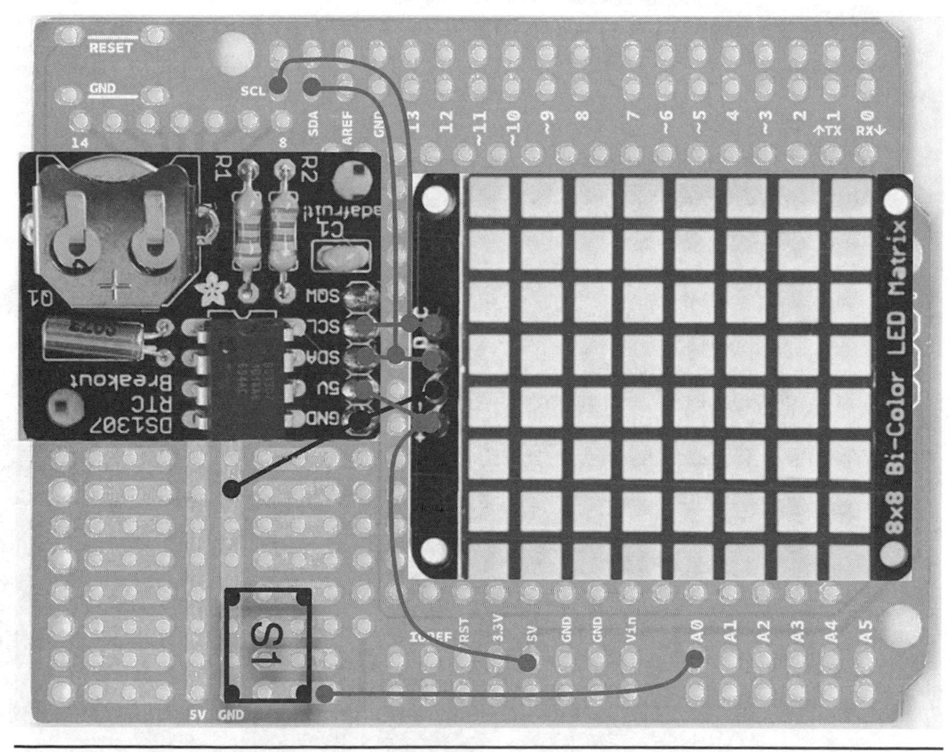

FIGURE 21-3 Protoshield layout for the clock.

Step 1: Assemble the RTC Module

The RTC module is supplied as a kit of parts, so the first step is to assemble the kit based on the instructions on the Adafruit website (www.ladyada.net/learn/breakoutplus/ds1307rtc.html).

Step 2: Assemble the LED Matrix Module

The LED matrix module is also supplied as a kit. The instructions for assembling this can be found at http://learn.adafruit.com/adafruit-led-backpack/bi-color-8x8-matrix.

Step 3:. Solder the Switch

Fit the switch into place as shown in Figure 21-3. Note that the switch must go with the wider gap between pins from top to bottom. That is, there should be two holes between the legs top to bottom and only one hole left to right. Figure 21-4 shows the board with the switch in place.

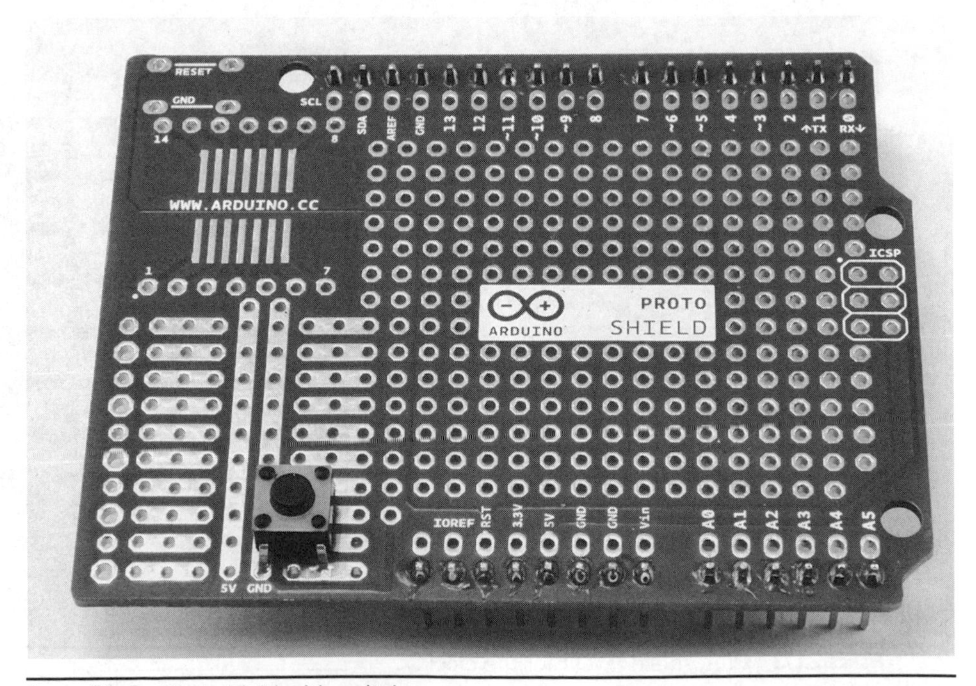

FIGURE 21-4 Protoshield with switch.

Step 4: Solder the Header Sockets

Cut the header socket into two lengths of four. Note that although the RTC module has five pins, we are only using four of them (5V, GND, SDA and SCL). Figure 21-5 shows the board with the header sockets in place. Note that you can dispense with these and solder the modules directly to the board if you wish. In Chapter 23, we will use exactly the same board layout but with a four-digit, seven-segment display instead of the LED matrix, so you may wish to use sockets so that you can easily swap the LED module over.

Step 5: Wire the Underside of the Protoshield

It now just remains to wire the underside of the board as shown in Figure 21-3. The underside of the completed board is shown in Figure 21-6.

You can now plug in the modules, making sure that the SQW pin of the RTC module is the pin left unconnected when you plug the module in.

FIGURE 21-5 Protoshield with sockets.

FIGURE 21-6 Underside of the Protoshield.

Software

The modules used in this project both have Arduino libraries that need to be installed in your Arduino IDE before the sketches will work. To install the libraries for the LED module, go to https://github.com/adafruit/Adafruit-LED-Backpack-Library and select the option to download as a zip. The icon on the webpage is labeled "Zip" and looks like a cloud with an arrow coming out of it.

The zip file is small, so it will not take long to download. When it has downloaded, unzip the folder and rename it from "Adafruit-LED-Backpack-Library-master" to "AdafruitLEDBackpack," and then copy it to your Arduino "Libraries" folder. You will need to do the same for the Adafruit_GFX library, which you can find at https://github.com/adafruit/Adafruit-GFX-Library.

If this is the first time you have added a library, you will need to create a "Libraries" folder by opening the "Arduino" folder in your "Documents" folder and then creating a new folder called "Libraries." Place the downloaded and renamed file into the "Libraries" folder. Note that for the Arduino IDE to be able to use the newly installed library, you will need to exit it and start it again. Don't do that yet, however, because you have two more libraries to install.

Follow the same procedure downloading the library from https://github.com/adafruit/Adafruit-GFX-Library. Rename the unzipped folder from "Adafruit-GFX-Library-master" to "AdafruitGFXLibrary," and copy it into the "Libraries" folder.

There is one more library to install for the RTC module from https://github.com/adafruit/RTClib. Once again, rename the unzipped folder from "RTClib-master" to "RTClib" and move it into the "Libraries" folder. Now restart the Arduino IDE so that it recognizes the new libraries. All the programs (or *sketches* as they are called in the Arduino world) are also available as a download from the author's website at www.simonmonk.org.

Here is the sketch for the project (Ch_21_LED_matrix_clock):

```
#include <Wire.h>
#include <RTClib.h>
#include <Adafruit_LEDBackpack.h>
#include <Adafruit_GFX.h>

int buttonPin = A0;

RTC_DS1307 RTC;
Adafruit_BicolorMatrix matrix = Adafruit_BicolorMatrix();

char* monthNames[] = {"Jan", "Feb", "Mar", "Apr", "May", "Jun",
  "Jul", "Aug", "Sep", "Oct", "Nov", "Dec"};


boolean showDate = false;

void setup()
{
  pinMode(buttonPin, INPUT_PULLUP);
  Wire.begin();
  if (! RTC.isrunning()
      || digitalRead(buttonPin) == LOW)
  {
    RTC.adjust(DateTime(__DATE__, __TIME__));
  }
  matrix.begin(0x70);
```

```
  matrix.setTextWrap(false);
  matrix.setTextSize(1);
}

void loop()
{
  if (showDate)
  {
    displayDate();
  }
  else
  {
    displayTime();
  }
}

void displayTime()
{
  DateTime now = RTC.now();
  for (int8_t x=7; x>=-36; x--)
  {
    checkButton();
    if (showDate)
    {
      delay(500);
      break;
    }
    matrix.clear();
    matrix.setCursor(x, 0);
    matrix.setTextColor(LED_GREEN);
    matrix.print(now.hour());
    matrix.setTextColor(LED_YELLOW);
    matrix.print(":");
    matrix.setTextColor(LED_GREEN);
    if (now.minute() < 10)
    {
      matrix.print("0");
    }
    matrix.print(now.minute());
    matrix.writeDisplay();
```

```
    delay(100);
  }
}


void displayDate()
{
  DateTime now = RTC.now();
  for (int8_t x=7; x>=-80; x--)
  {
    checkButton();
    if (! showDate)
    {
      delay(500);
      break;
    }
    matrix.clear();
    matrix.setCursor(x, 0);
    matrix.setTextColor(LED_RED);
    matrix.print(now.day());
    matrix.setTextColor(LED_YELLOW);
    matrix.print(" ");
    matrix.print(monthNames[now.month() -1]);
    matrix.setTextColor(LED_GREEN);
    matrix.print(" ");
    matrix.print(now.year());
    matrix.writeDisplay();
    delay(100);
  }
}


void checkButton()
{
  if (digitalRead(buttonPin) == LOW)
  {
    showDate = ! showDate;
  }
}
```

The first four lines include the libraries that we are going to use. The "Wire" library comes preinstalled with the Arduino IDE, and it is this that is used to communicate with the I2C serial interface.

Next, we define the button pin and create two variable called `RTC` and `matrix` that we will use to communicate with the RTC and LED matrix modules. The array of character strings `monthNames` is used later to display the names of the months when displaying the date. You can, if you wish, modify the names used here to, say, another language.

The Boolean variable `showDate` is used as a flag to indicate if it is the date or time that should be displayed. It is this variable that will be changed when the button is pressed.

The setup function has quite a bit to do. First, it sets up the button pin and then starts I2C communications going with `Wire.begin()`. It then checks to see if the RTC module is running. It will not be, the very first time it is used, because it only starts to run once the date and time have been set.

If they have not been set or the button is being held down, then the date and time are set to the special variables `__DATE__ and __TIME__`, which are the date and time at which the sketch is uploaded to the board. The LED matrix module is also initialized, and text wrapping turned off so that we can scroll the display.

The main loop function is very simple; it just decides if the date or time should be displayed and calls the appropriate function. The display functions `displayDate` and `displayTime` are very similar. After fetching the time from the RTC module, the `for` loop shifts the `x` position of the cursor to give the effect of scrolling. The screen is cleared, and then the message is displayed in a mixture of colors.

You will notice that inside the scrolling loop, there is also code to check whether the button is pressed so that if it is, the current message can be interrupted and the display mode toggled between date and time. In `displayTime`, if the minute is less than `10`, then it is padded with an extra zero.

Summary

In Chapter 22, we will use the RTC module again, but this time, to make a binary clock.

CHAPTER 22

Binary Clock

Difficulty: ★★	Cost guide: $20

This project uses different colored LEDs to display the time. However, it displays the time in binary, with a 1 being indicated by the LED being lit (Figure 22-1).

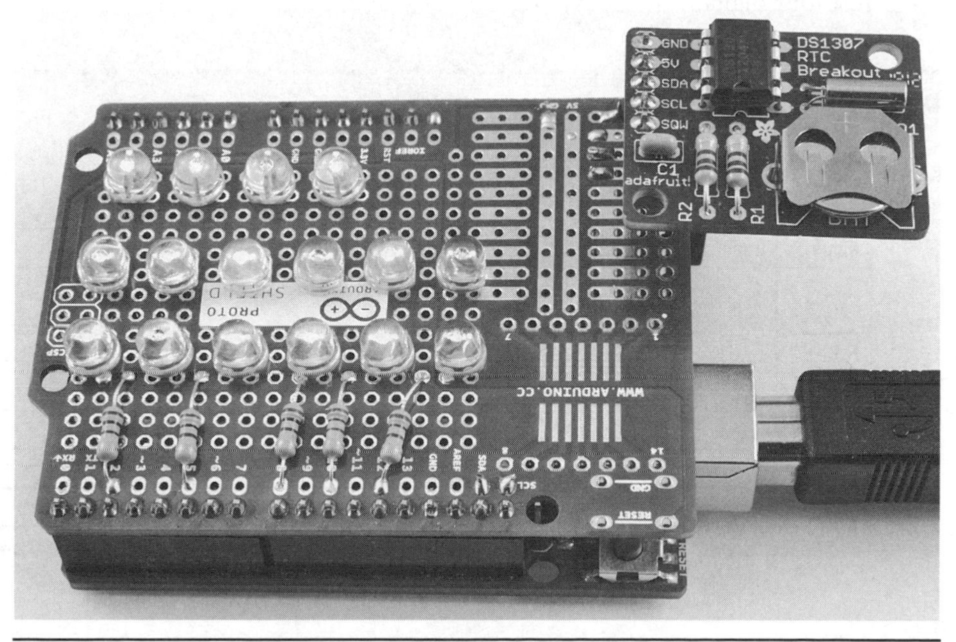

FIGURE 22-1 Binary clock.

Binary

The decimal numbers that we usually use have 10 possible digits in each position. That is, each digit is between 0 and 9, and if we need a number bigger than 9, we add another digit that counts in 10s rather than single units.

Binary only has two possible values in each digit position—a 1 or a 0. Therefore, this means that if we want to represent a number greater than 1, we have to add another digit. Counting in binary from 0 to 10 goes like this:

```
0, 1, 10, 11, 100, 101, 110, 111, 1000, 1001, 1010
```

The decimal values of each digit for a four-digit binary number are 8, 4, 2, and 1.

To convert binary into the more familiar decimal, add up all the decimal equivalents of that digit where there is a 1 in that position. For example, binary 1101 is 8 + 4 + 1 = 13 in decimal, and 0011 binary is 2 + 1 = 3 in decimal.

To represent the time, we will display the hours, minutes, and seconds, each as a separate binary number. With six binary digits, you can represent any number between 0 and 63, so six digits will be fine to represent the minutes and seconds. To display the hour, we will use 1 to 12 rather than a 24-hour clock, so we need just four digits.

Parts

To build this project, you will need the following:

Name	Quantity	Description	Appendix
	1	Protoshield printed circuit board (PCB) and header pins	A3, H1
R1–5	5	100 Ω resistors	R3
	20	LEDs (any color)	S1–5
	1	Adafruit real-time clock (RTC) breakout	M5
		Solid-core wire	
	1	Four-way header socket[a]	H2

[a]*The header socket is optional; you can solder the module directly to the Protoshield if you prefer.*

Protoshield Layout

This project uses a very cunning technique called *charlieplexing*. This allows all 20 LEDs to be controlled by just five Arduino pins. What's more, it only requires five series resistors and is a great deal easier to wire than if we were connecting each LED to a separate Arduino pin. The name *charlieplexing* comes from the inventor, Charlie Allen, at a company called *Maxim*, and it takes advantage of the feature of Arduino input-output (I/O) pins that allows them to be changed from outputs to inputs while a sketch is running. Figure 22-2 shows the arrangement for controlling six LEDs with three pins. Table 22-1 shows how the pins should be set to light a particular LED.

TABLE 22-1 Charlieplexing with Three Pins

LED	Pin 1	Pin 2	Pin 3
A	High	Low	Input
B	Low	High	Input
C	Input	High	Low
D	Input	Low	High
E	High	Input	Low
F	Low	Input	High

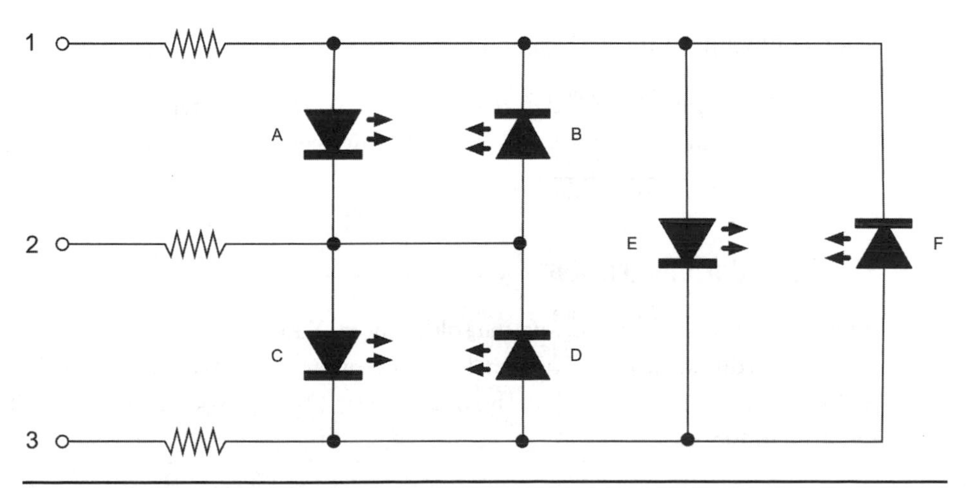

FIGURE 22-2 Charlieplexing.

The number of LEDs that can be controlled per Arduino pin is given by the formula

$$\text{LEDs} = n \times n - n$$

Thus, if we use 4 pins, we can have 16 – 4 = 12 LEDs, and 10 pins would give us a massive 90 LEDs and an awful lot of wiring to do. In this case, we need 5 I/O pins, which would give us a maximum of 20 pins, but we only need 16.

To display more than one LED at a time, you have to use the same kind of refreshing technique that we did with the persistence-of-vision and LED cube projects to give the illusion that more than one LED is lit. Figure 22-3 shows the schematic diagram for the project.

The LEDs are labeled with the letters *h*, *m*, and *s* for "hour," "minute," and "second," respectively, and a digit, where 0 is the least significant binary digit (decimal value 1) and 5 is the most significant (decimal 32). I used all blue LEDs—it's what I had available. But there is no reason why you shouldn't use different colors. However, it is worth trying to select LEDs of similar brightness.

Construction

Figure 22-4 shows the Protoshield layout. The junctions up near the resistors have quite a lot of connections, all coming into one place, so check carefully that all the connections you need have been made. As always, start by attaching header pins to the Protoshield.

Step 1: Assemble the RTC Module

The RTC module is supplied as a kit of parts, so the first step is to assemble the kit based on the instructions on the Adafruit website (www.ladyada.net/learn/breakoutplus/ds1307rtc.html).

Step 2: Solder the Resistors

Fit all the resistors in place, and the solder them. You can trim off the excess leads near the Arduino header pins, but leave the other ends. You will need them for making connections under the board. Figure 22-5 shows the board with the resistors in place.

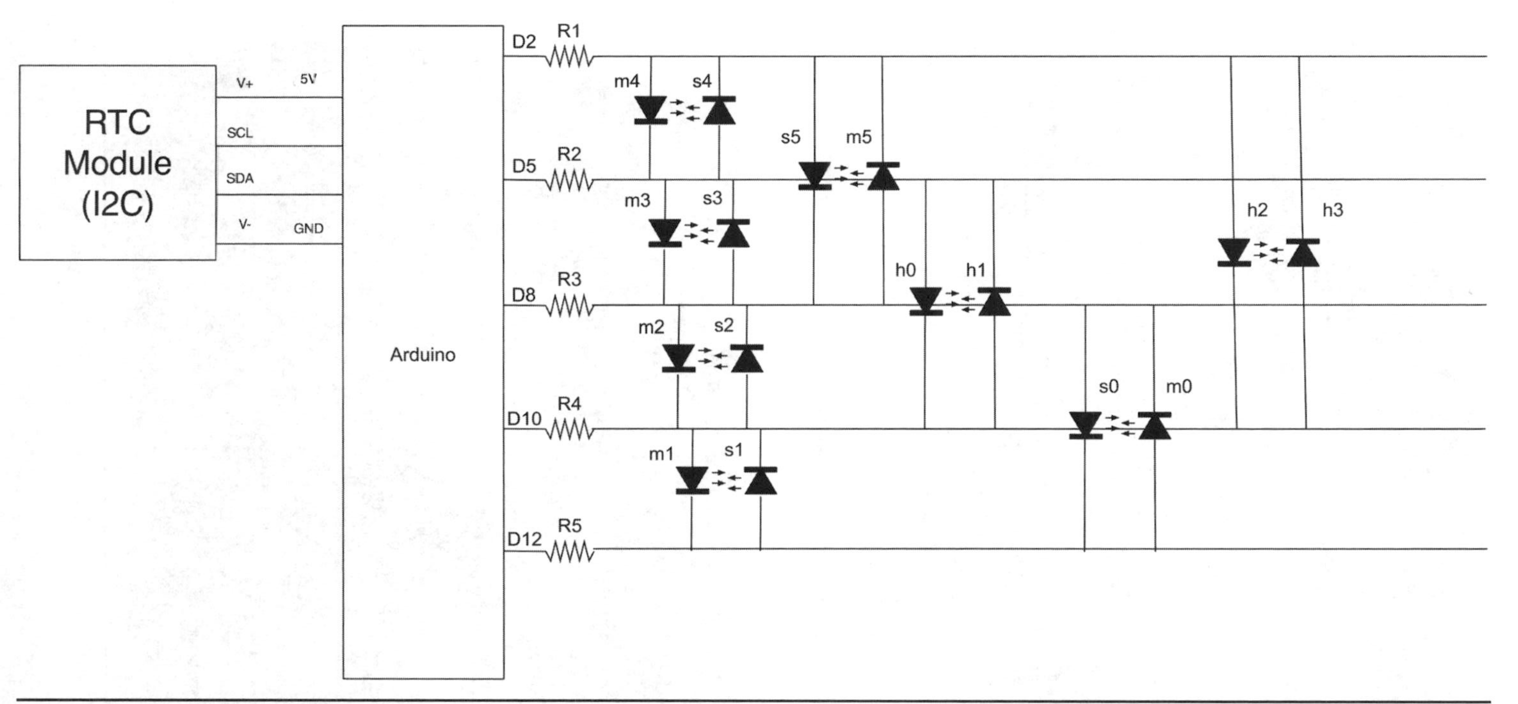

FIGURE 22-3 Schematic diagram for the binary clock.

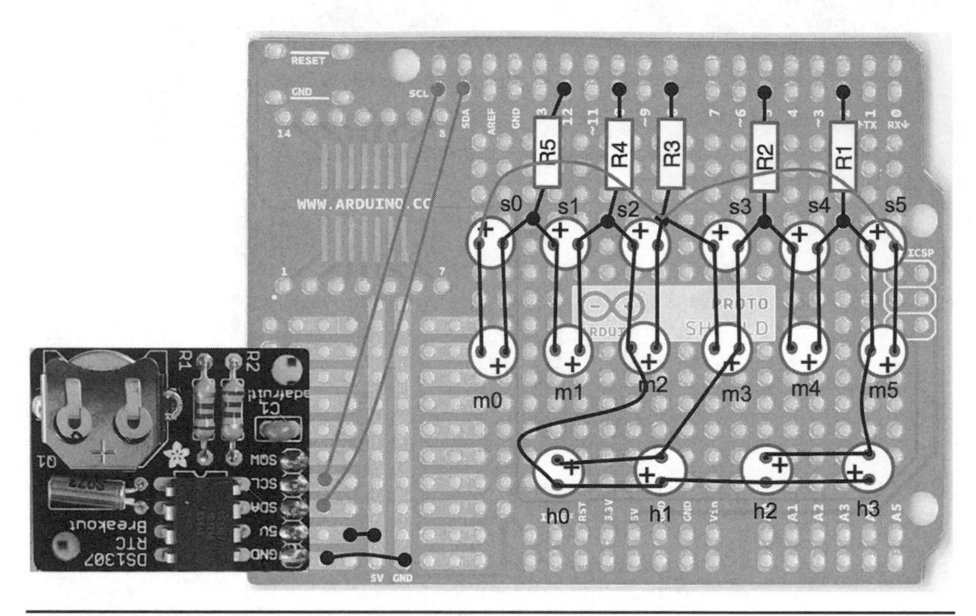

FIGURE 22-4 Protoshield layout for the binary clock.

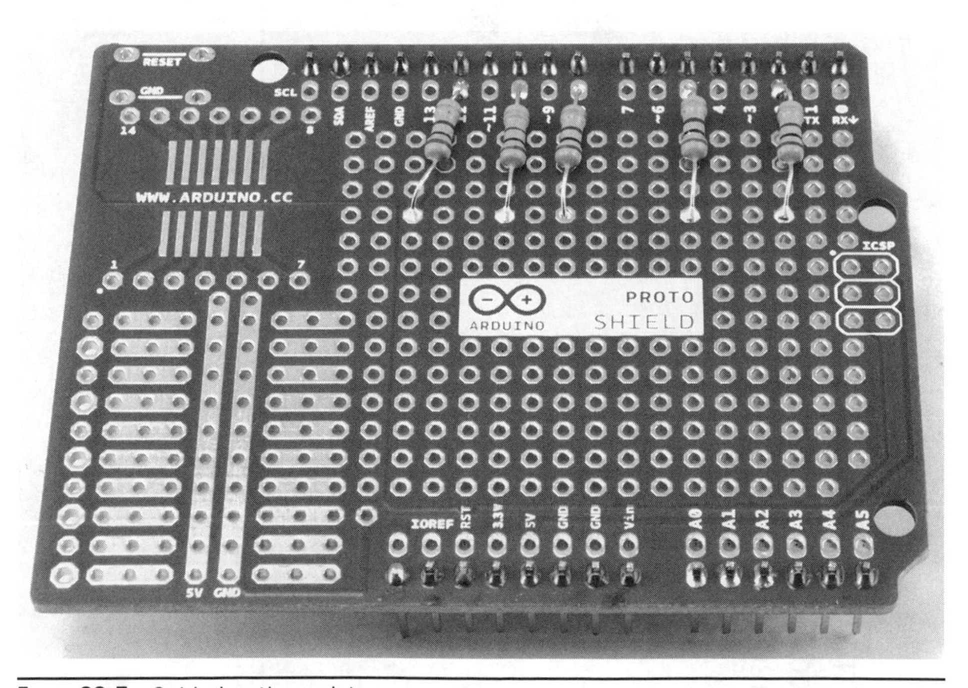

FIGURE 22-5 Soldering the resistors.

Step 2: Solder the LEDs

It is critical that you get the LEDs the right way around because it will be difficult to find out if one is the wrong way around once the leads are cut off. Remember that the longer lead is the positive lead. Check every LED against Figure 22-4 before you solder it. Figure 22-6 shows the Protoshield with the LEDs attached.

Step 3: Solder the Header Socket or RTC Module

Soldering the RTC module directly to the board will lower its profile, making the LEDs easier to see and the board simpler to mount in a box. However, it does make it harder if you want to use the RTC module in a different project. With the header socket in place, your board should look like Figure 22-7.

Step 4: Wire the Underside of the Protoshield

Figure 22-8 shows the wiring diagram from the point of view of the underside of the board. With all the components in place, the underside of the board will look

Figure 22-6 Soldering the LEDs.

FIGURE 22-7 Soldering the header.

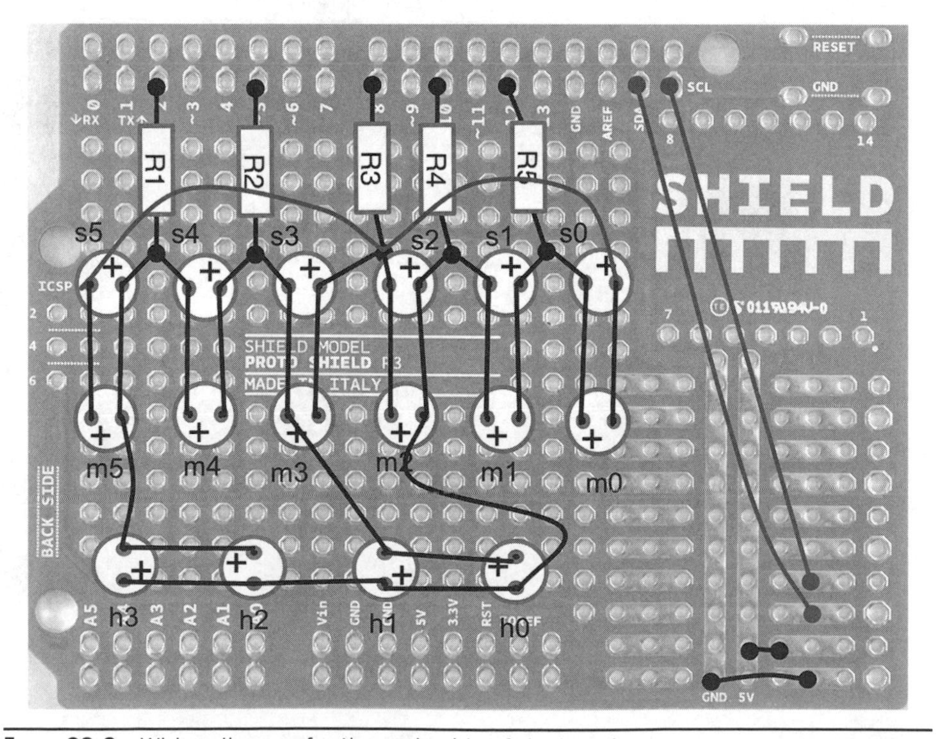

FIGURE 22-8 Wiring diagram for the underside of the board.

a bit like a porcupine. You can tidy things a bit by connecting the LED leads that belong in pairs as shown in Figure 22-9.

Using Figure 22-8, you can now start soldering and clipping. If you do cut off a lead that you later find you needed, you can always rejoin it with a bit of snipped resistor lead. When all the connections have been made using the component leads, the board will look like Figure 22-10.

Finally, use some solid-core wire for the remaining connections that cannot be made just by bending the component leads. The underside of the completed board is shown in Figure 22-11, and the finished board is shown in Figure 22-12.

Figure 22-9 Underside of the board.

FIGURE 22-10 Underside of the board, soldered.

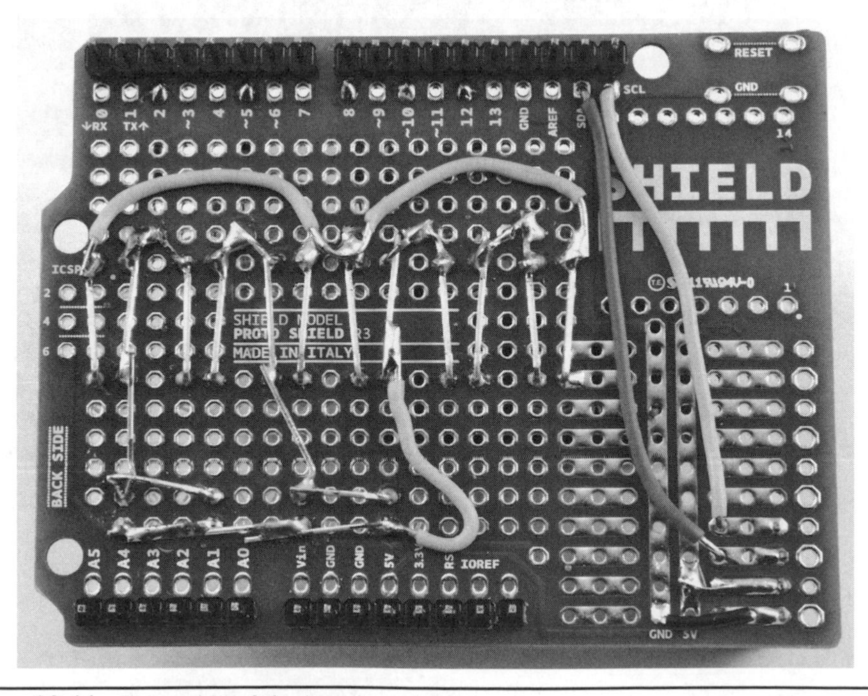

FIGURE 22-11 Underside of the board, complete.

FIGURE 22-12 Finished board.

Software

The software (ch_22_binary_clock) needs the same RTC module library that you first used back in Chapter 21. It also needs a new library that we have not used before called *TimerOne*. This library can be downloaded from https://code.google.com/p/arduino-timerone/downloads/list.

You will need to unzip the folder, rename it to be just "TimerOne," and move it into the "Libraries"folder inside your "Arduino" folder. You also will need to restart the Arduino IDE for the library to be picked up. All the programs (or sketches as they are called in the Arduino world) are also available as a download from the author's website at www.simonmonk.org.

Here is the sketch:

```
// Chapter 22. Binary Clock
#include <Wire.h>
#include <RTClib.h>
#include <TimerOne.h>

byte numPins = 5;
byte ledPins[] = {2, 5, 8, 10, 12};

byte hourLeds[][2] = {{1, 3}, {3, 1}, {0, 3}, {3, 0}};
byte minLeds[][2] = {{4, 2}, {3, 4}, {2, 3},
                {1, 2}, {0, 1}, {2, 0}};
```

```
byte secLeds[][2] = {{2, 4}, {4, 3},
                {3, 2}, {2, 1}, {1, 0}, {0, 2}};

RTC_DS1307 RTC;

int hour, min, sec;

void setup()
{
  Wire.begin();
  if (! RTC.isrunning())
  {
    RTC.adjust(DateTime(__DATE__, __TIME__));
  }
  Timer1.initialize(20000); // uS
  Timer1.attachInterrupt( refresh );
}

void loop()
{
  static long lastTick;
  long thisTick = millis();
  if (thisTick > lastTick + 500l)
                      // every 0.5 seconds
  {
    DateTime now = RTC.now();
    hour = now.hour();
    min = now.minute();
    sec = now.second();
    lastTick = thisTick;
  }
  refresh();
}

void refresh()
{
  int s = sec;
  int m = min;
  int h = hour;
```

```
  for (int bit = 0; bit < 6; bit++)
  {
    if (s & 1)
    {
      setLed(secLeds[bit]);
    }
    s = s >> 1;
    if (m & 1)
    {
      setLed(minLeds[bit]);
    }
    m = m >> 1;
    if (h & 1 && bit < 4)
    {
      setLed(hourLeds[bit]);
    }
    h = h >> 1;
   }
   allOff(); // so last led not brighter
}

void setLed(byte pins[])
{
  byte plusPin = ledPins[pins[0]];
  byte minusPin = ledPins[pins[1]];
  allOff();
  pinMode(plusPin, OUTPUT);
  digitalWrite(plusPin, HIGH);
  pinMode(minusPin, OUTPUT);
  digitalWrite(plusPin, LOW);
}

void allOff()
{
  for (byte pin = 0; pin < numPins; pin++)
  {
    pinMode(ledPins[pin], INPUT);
  }
}
```

This sketch is largely driven through the use of arrays. There is an array for the pins used for the LEDs (`ledPins`). Then there are three more arrays, one for each of the hour minute, and second LEDs. The elements of each of these LEDs are themselves an array with just two elements, the first of which is the `HIGH` pin for the LED in question and the second of which is the `LOW` pin.

To keep the refresh as fast and consistent as possible, the "TimerOne" library is used to link the `refresh` function with a timer-driven interrupt. This means that the `refresh` function will be called every 20 milliseconds, interrupting whatever else the Arduino is doing.

The main `loop` function is therefore just responsible for reading the time from the RTC module and putting the result into three variables: `hours`, `minutes`, and `seconds`. The RTC module is not checked every time around the loop but just every half second. This also helps to keep the LEDs lit more of the time.

The charlieplexing code lives in the functions `setLed` and `allOff`. The `allOff` function simply sets all the LED control pins to be inputs. The `setLed` function calls this and then finds which pins for that LED are the positive and negative pins and sets them to be `HIGH` and `LOW` outputs, respectively.

Summary

In Chapter 23, we will build yet another clock, but this time using a seven-segment LED display.

CHAPTER 23

Seven-Segment LED Clock

Difficulty: ★	Cost guide: $25

This project uses the same real-time clock (RTC) module used in Chapters 21 and 22 but this time with a more conventional seven-segment LED display (Figure 23-1).

When the button is pressed, the clock toggles between displaying the time in hours and minutes and displaying the seconds. The time is set automatically when uploading the sketch if the button is pressed during the upload.

Parts

To build this project you, will need the following:

Name	Quantity	Description	Appendix
	1	Protoshield printed circuit board (PCB) and header pins	A3, H1
		Adafruit four-digit, seven-segment LED module	M6
	1	Adafruit RTC breakout	M5
		Solid-core wire	
S1	1	Click switch	C2
	2	Four-way header sockets[a]	H2

[a] *The header socket is optional; you can solder the modules directly to the Protoshield if you prefer.*

Figure 23-1 Seven-segment LED clock.

Protoshield Layout

Figure 23-2 shows the schematic diagram for the project. This is exactly the same as the LED matrix project from Chapter 21, and the seven-segment display has the same four connections as the matrix display. Thus, if you built Chapter 21 with sockets, you can just unplug the matrix display and plug in the seven-segment module.

Construction

If you made the project in Chapter 21, then there is not really anything to make here. If you didn't, then head back to Chapter 21 and follow the instructions there.

Software

The sketch for this project (ch_23_7_seg_clock) is also similar to that of Chapter 21. It also requires the same libraries to be installed into your Arduino IDE. See Chapter 21 for installing the libraries into the Arduino IDE.

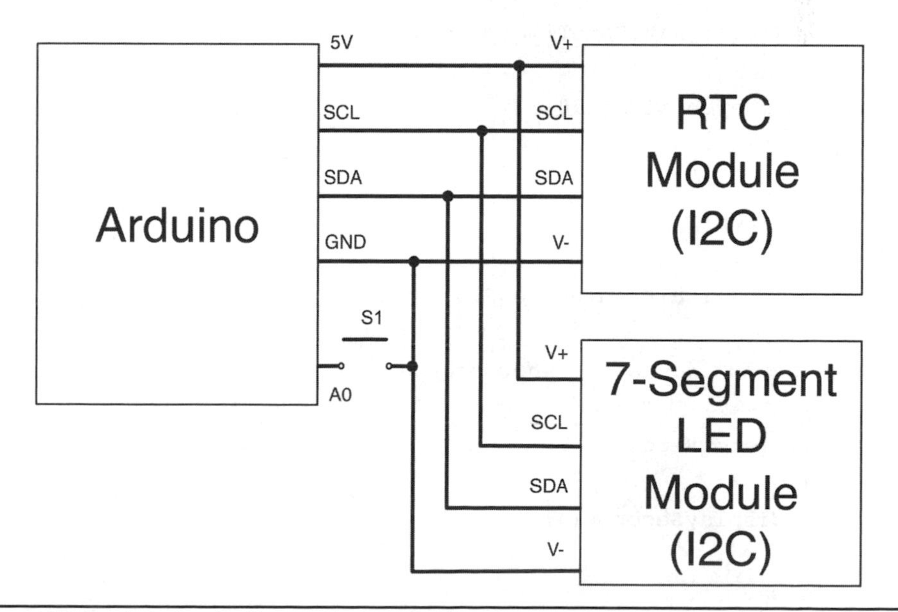

FIGURE 23-2 Schematic diagram for the clock.

All the programs (or sketches as they are called in the Arduino world) are also available as a download from the author's website at www.simonmonk.org.

The sketch is actually a bit simpler than that of the project in Chapter 21:

```
#include <Wire.h>
#include <RTClib.h>
#include <Adafruit_LEDBackpack.h>
#include <Adafruit_GFX.h>

int buttonPin = A0;

RTC_DS1307 RTC;
Adafruit_7segment display = Adafruit_7segment();

boolean showSeconds = false;

void setup()
{
  pinMode(buttonPin, INPUT_PULLUP);
  Wire.begin();
  if (! RTC.isrunning()
     || digitalRead(buttonPin) == LOW)
  {
    RTC.adjust(DateTime(__DATE__, __TIME__));
  }
  display.begin(0x70);
}

void loop()
{
  if (digitalRead(buttonPin) == LOW)
  {
    showSeconds = ! showSeconds;
  }
  if (showSeconds)
  {
    displaySeconds();
  }
  else
  {
```

```
    displayTime();
  }
}

void displayTime()
{
  static boolean colonState;
  DateTime now = RTC.now();
  int h = now.hour();
  int m = now.minute();
  display.print(h * 100 + m);
  display.drawColon(colonState);
  colonState = ! colonState;
  display.writeDisplay();
  delay(500);
}

void displaySeconds()
{
  DateTime now = RTC.now();
  int s = now.second();
  display.clear();
  display.writeDigitNum(3, s / 10);
  display.writeDigitNum(4, s % 10);
  display.writeDisplay();
  delay(500);
}
```

Much of the code is the same as in the earlier project, so please look at the description in Chapter 21 if there is anything you do not understand.

The `displayTime` function fetches the hour and minute separately, multiples the hour by 100, and adds the minute to construct a three- or four-digit number to represent the time. The flashing colon effect when displaying the time is achieved by using a static variable (`colonState`) that remembers its value between successive calls of the function. This is toggled every time the time is displayed, and because there is a half-second delay at the end of the time display, this has the effect of making the colon blink roughly once a second.

To display the seconds, the display is first cleared, and then the two digits of the second are written individually to the two last segments. Note that these are segments 3 and 4 rather than the 2 and 3 that you might be expecting. This is so because segment 2 is actually the colon. It's a bit strange, but this is just how it is wired.

Summary

In Chapter 24, we take a break from digital clocks and hack a good old-fashioned analog clock.

CHAPTER 24
Hacked Analog Clock

Difficulty: ★★★★	Cost guide: $10

Analog wall clocks normally use a quartz crystal and some electronics to generate pulses that drive a tiny actuator that ticks the clock on by a second. The movement of the minute and hour hands uses gears to make them turn at the correct rate. With a bit of care, you can hack the clock so that the pulses are generated by an Arduino (Figure 24-1).

This idea was introduced to me by Esa Pollari. Many thanks to him for developing this excellent idea.

Parts

This project is compatible with both the Arduinos Uno and Leonardo. To build this project, you will need the following:

Part	Quantity	Description	Appendix
	1	270 Ω resistor	R1
	1	Header pins (two way)	H2
	1	Low-cost quartz wall clock	
		Thin multicore wire	

This project is somewhat experimental. This approach has worked with the two or three different types of wall clocks that I have tried it on but will no doubt not work quite the same on all quartz wall clocks. There is also a good chance that

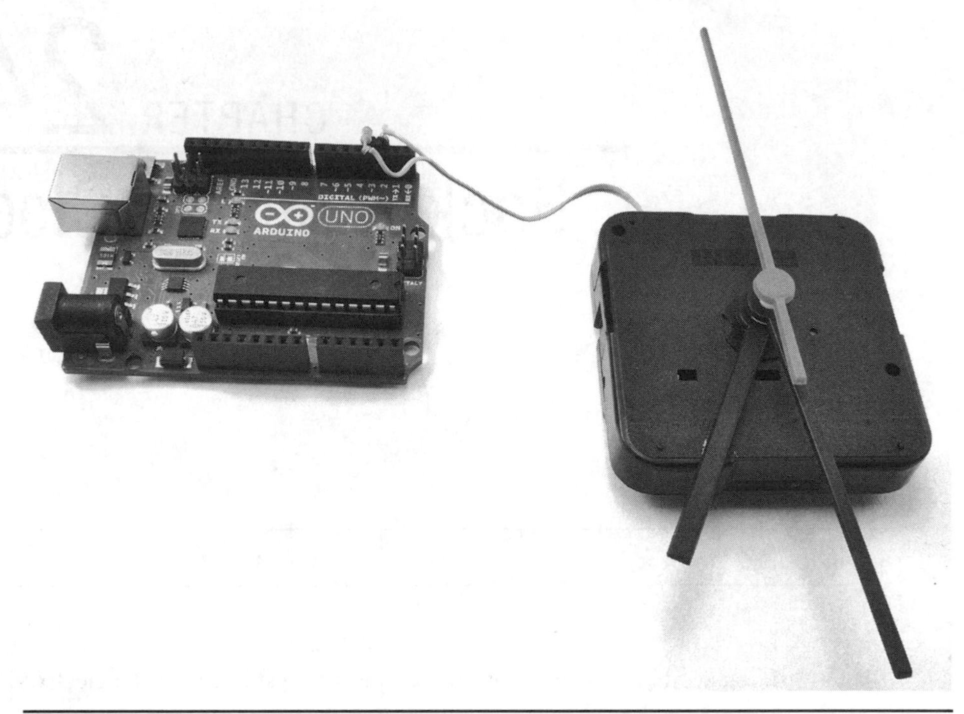

FIGURE 24-1 Hacked analog clock.

when you take the clock mechanism apart, all the tiny cogs will fall out, and that will be the end of the clock. Therefore, do not choose anything but the cheapest of clocks to try this with—at least for the first time. Figure 24-2 shows the clock that I used, which cost less than $5.

You are going to need some thin multicore wire (just a few inches) that will be connected inside the clock mechanism and fed out through the clock's case. I used wire reclaimed from an old computer lead.

Construction

There is not very much construction to do, and if you are very careful (or lucky), then the clock mechanism will stay in place. If you are less careful, then the most complicated part of the project will be putting the clock back together again.

Step 1: Remove the Clock Mechanism

We need to take the clock mechanism off the body of the clock. Many cheap clocks just clip together, making this quite easy. Start by removing the front glass

FIGURE 24-2 Low-cost wall clock.

(or more likely plastic), and then carefully pull off the hands. At this point, you should be able to unclip the mechanism from the back of the clock.

Step 2: Open the Mechanism Enclosure

This is the step where things can go horribly wrong. Look carefully at the clock mechanism, and work out where any clips or screws are that will hold the two parts of the case together. The aim here is to open the case in such a way that all the cogs stay in the case but that the tiny PCB with the control electronics is exposed so that we can hack it.

Thus, carefully lever open the clips or remove the screws and keeping everything horizontal. Try and see into the case enough to see which way up you need it to be to get at the PCB (Figure 24-3). Flip the case over if you need to, and then pull the two sections apart (Figure 24-4).

Figure 24-3 Carefully open the clock case.

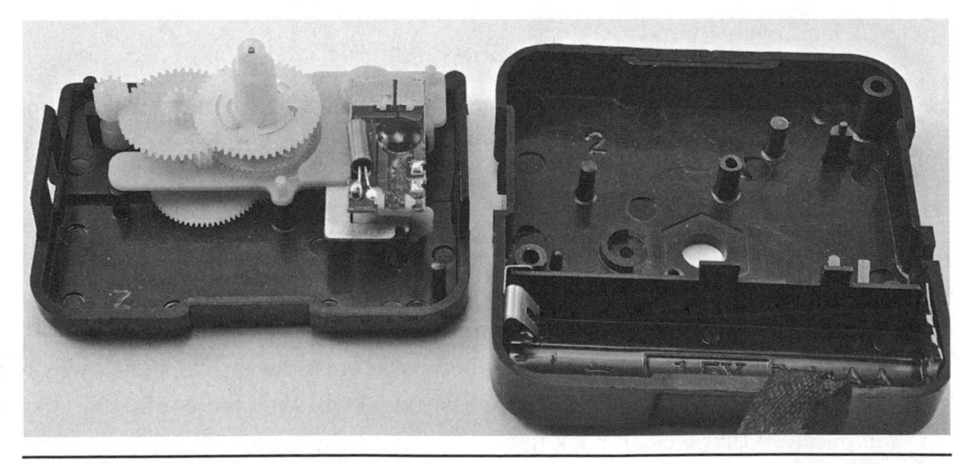

Figure 24-4 Clock mechanism exposed.

Step 3: Cut the PCB Tracks

If you look carefully at the PCB, you can see that the wires from the coil attach to the two solder pads at the front of the PCB (these are circled in Figure 24-5).

These two connections are all the Arduino needs to control the clock. It effectively replaces everything else on the PCB. However, we do not want the remainder of the electronics to interfere with the pulses generated by the Arduino, so we are going to sever the tracks that lead from these two pads.

The copper tracks on the PCB will be very thin and can be easily broken by scraping them repeatedly with a sharp screwdriver blade. Figure 24-6 shows the PCB with the two tracks cut just before they disappear under the blob of plastic that shields the clock control chip.

You only actually need to cut one of the tracks, but here both have been cut. The cut tracks are circled in Figure 24-6.

Step 4: Prepare the Lead

We are going to use a few inches of two-way lead to take the connections to the clock coil outside the clock mechanism's enclosure. Before we attach the lead, let's

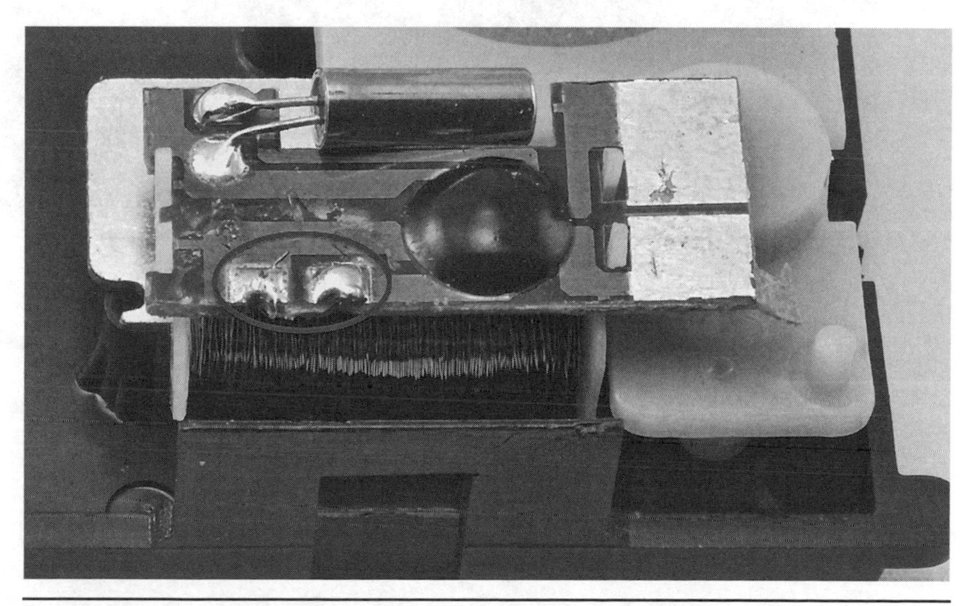

FIGURE 24-5 The clock's PCB.

prepare it. Figure 24-7 shows the end of the lead that is going to plug into the Arduino.

You can see that the leads of the 270 Ω resistor have been shortened and that the resistor is attached in-line with the lead. The other end of the lead just needs both wires to be stripped and tinned with solder.

Step 5: Solder the Leads to the Clock Coil

You now need to solder the tinned wires of the lead to the pads on the clock's PCB, as shown in Figure 24-8.

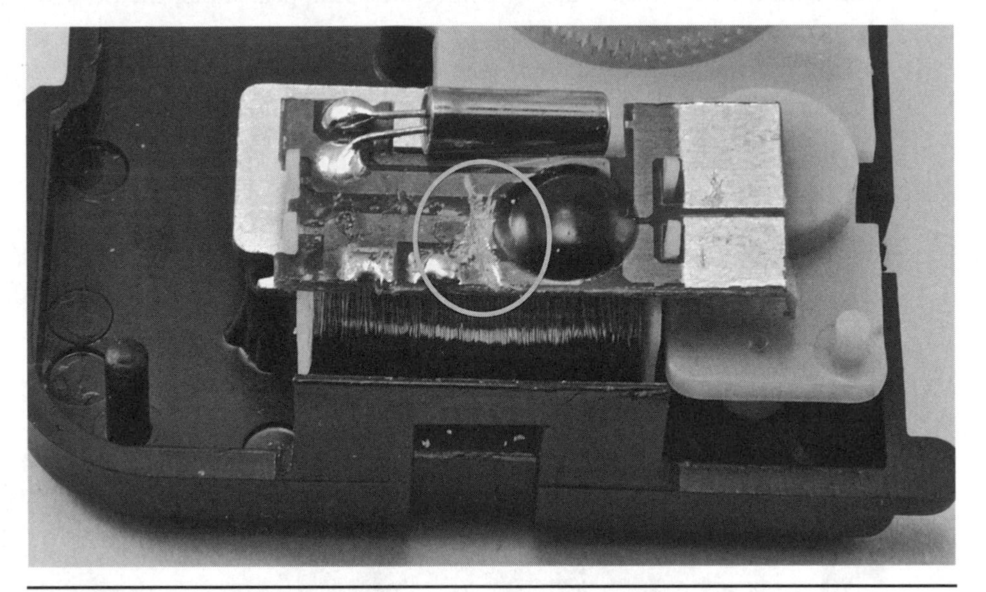

FIGURE 24-6 Cutting the PCB tracks.

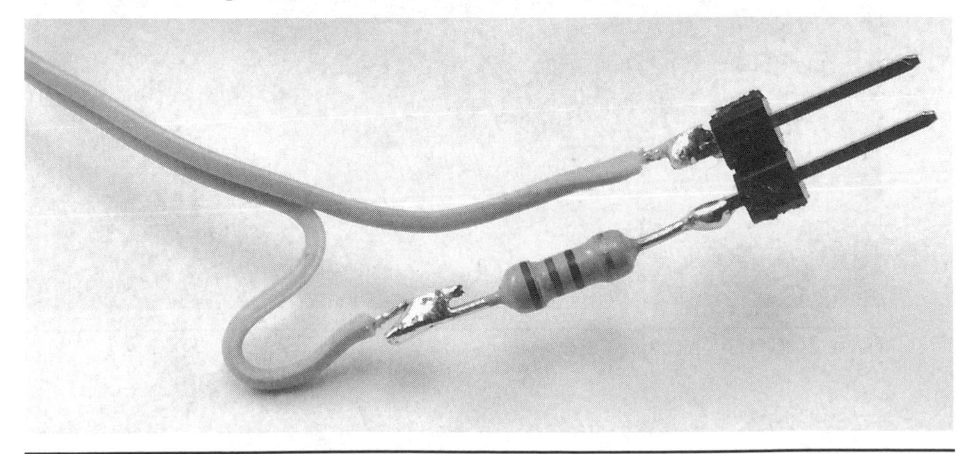

FIGURE 24-7 Preparing the lead.

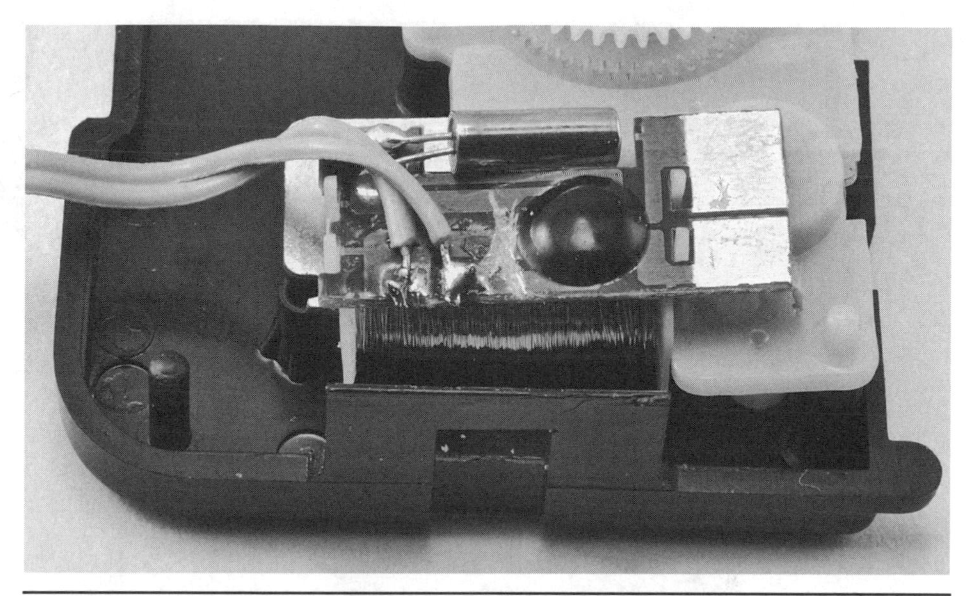

FIGURE 24-8 Soldering the leads to the clock PCB.

Step 6: Reassemble the Clock Mechanism Case

All that remains of the construction is to reassemble the clock case, allowing the lead to escape through one side of the enclosure. Fortunately, plastic will bend, allowing the lead to leave the case without the need to drill a hole for it. Figure 24-9 shows the reassembled clock mechanism.

Software

There are two sketches that accompany this project. The first to download and install is a test sketch (ch_24_hacked_wall_clock_test). This sketch will help you to find the right pulse width for the clock mechanism. Both sketches for this project need the "Timer" library, which can be downloaded and installed from http://playground.arduino.cc/Code/Timer.

Plug the pin header on the end of the lead from the clock mechanism into pins 3 and 3 of the Arduino. It does not matter which way around. You also will need to attach the second hand so that you can see the clock operating.

Upload the sketch ch_24_hacked_wall_clock_test, and open the serial monitor (Figure 24-10). You will see a prompt asking you to enter a pulse length. The pulse lengths will vary between clock mechanisms. So you may be lucky and find that the default pulse length of 30 milliseconds works just fine. If you find that the

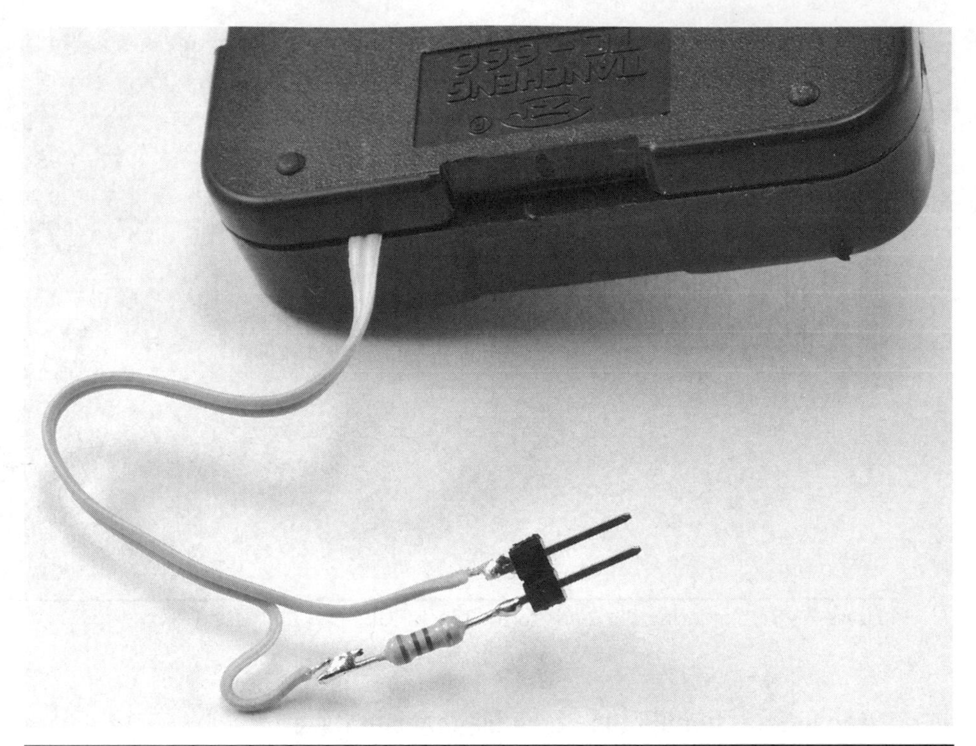

FIGURE 24-9 The hacked clock mechanism.

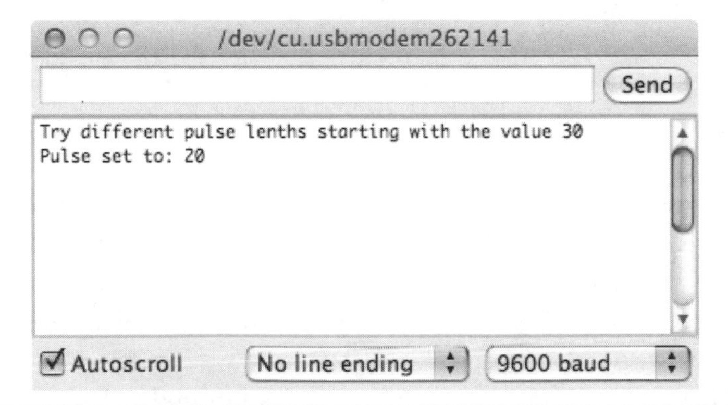

FIGURE 24-10 Experimenting with the pulse length.

second hand just jitters or does not tick over every time, then use trial and error to work out the best pulse length. Make a note of the number because you will now need to us it in the full sketch for the project.

The sketch starts with the definition of two constants for the two pins that are to be connected to the coils of the clock mechanism and the `pulseLen` variable.

```
const int pinA = 2;
const int pinB = 3;
int pulseLen = 30;
```

The `setup` function defines these pins to be outputs and starts serial communication. It then provides the user with a prompt of what he or she should send from the serial monitor. Finally, `setup` starts a timer that will run the function `pulse` every second.

The `loop` function checks to see whether a serial message has been received, and if it has, it reads it as a number and sets the variable `pulseLen` to this value before sending a confirmation message back to the serial monitor.

```
void loop()
{
  if (Serial.available())
  {
    pulseLen = Serial.parseInt();
    Serial.print("Pulse set to: ");
    Serial.println(pulseLen);
  }
  t.update();
}
```

The call to `t.update()` allows the timer to check whether a second has elapsed since the last call to `pulse`.

There are two versions of the `pulse` function. The first one has no parameters and calls the second, which takes two pins as its parameters. Both are as follows:

```
void pulse()
{
  static int state;
  pulse(state, !state);
  state = ! state;
}
```

```
void pulse(int a, int b)
{
  digitalWrite(pinA, a);
  digitalWrite(pinB, b);
  delay(pulseLen);
  digitalWrite(pinA, LOW);
  digitalWrite(pinB, LOW);
}
```

The motor requires the polarity of the voltage across it to be reversed each time there is a tick. The first `pulse` function takes care of this by toggling the variable `state` each time it is called. The variable `state` is static, which means that its value is associated with the `pulse` function but will not be lost between successive invocations of `pulse`. The second `pulse` function sets the two pins at opposite poles, holds them there for the necessary pulse length, and then turns both pins off again, ready for the reverse-polarity pulse that will be sent the next time `pulse` is called.

Using the Project

The main sketch for this project is ch_24_hacked_wall_clock. You can now upload this, and you should find the second hand moving once per second. You will need to change the value of the constant `pulseLen` to match the value that you found with the test sketch. This sketch will allow you to change the speed of the clock (dramatically) by sending a number between 0 and 9 from the serial monitor (Figure 24-11). You could, for example, run leads to the clock from

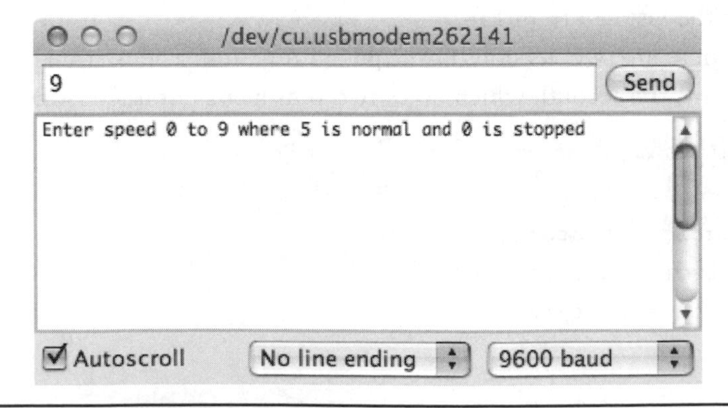

FIGURE 24-11 Controlling the speed of the clock.

your Arduino and then use the serial monitor to change the speed at which the clock runs.

Summary

In this chapter, we have established the basics of using an analog clock with an Arduino. In Chapter 25, we will modify three such clocks, driving them all from one Arduino to make a "world clock" that shows the time in three locations.

CHAPTER 25
World Clock

Difficulty: ★★★★	Cost guide: $20

In Chapter 24, you found out how to hack a wall clock so that you could control the speed of the hands using an Arduino. In this chapter, we will use an Arduino to control three such hacked clocks and make a *world clock* (Figure 25-1).

Because the Arduino will be generating the pulses for the clocks, all the clocks will stay in sync.

Parts

This project is compatible with both the Arduinos Uno and Leonardo. To build this project, you will need the following:

Part	Quantity	Description	Appendix
	3	270 Ω resistors	R1
	1	Header pins (six way)	H2
	3	Low-cost quartz wall clocks	
	1	Tea tray	
		Thin multicore wire	

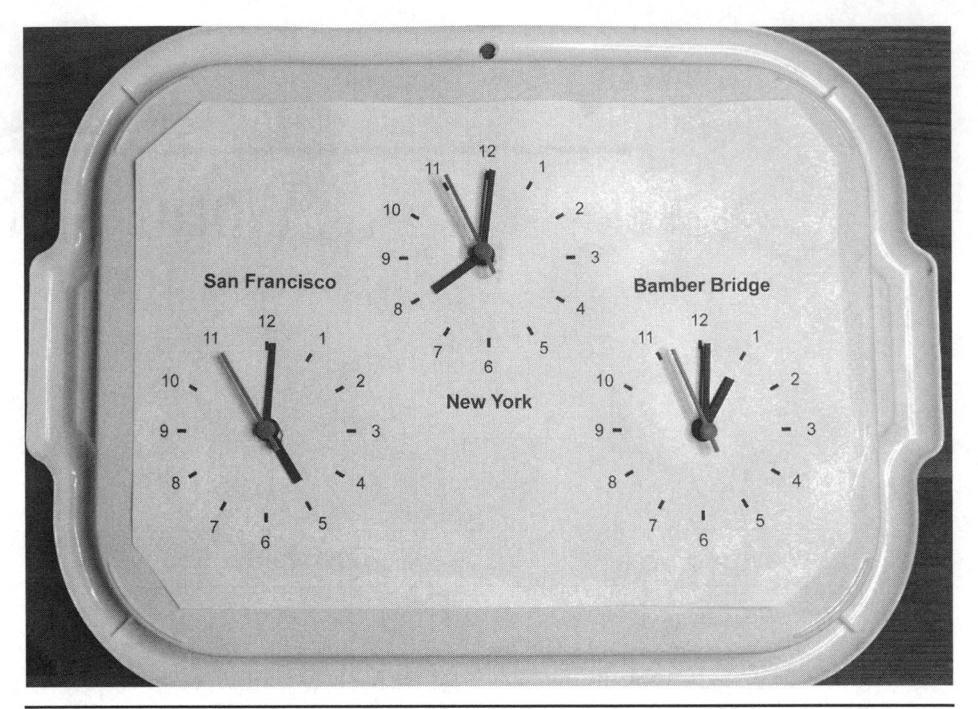

FIGURE 25-1 World clock.

Construction

As you can see from Figure 25-2, the clock is simply three of the hacked clock mechanisms, and the mechanisms are mounted on a plastic tea tray. The Arduino is actually an older model that predates the Uno, but an Uno or a Leonardo will work just fine.

Step 1: Make Three Hacked Clock Mechanisms

The first step is to follow the instructions in Chapter 24 three times to make three hacked clocks. Rather than have three separate two-pin header plugs, you can just solder all six connections to the header pins because we are going to use Arduino pins D2 to D7, which are all in a row.

Step 2: Drill the Tray and Glue the Mechanisms in Place

I used a hot-glue gun to fix the clock mechanisms in place behind holes drilled for the shafts.

FIGURE 25-2 Construction of the world clock.

Step 3: Print a Clock Face

The clock hands by themselves are of little use, so fire up your favorite drawing software (or pen and paper) and make a set of clock faces for the world clock.

Software

The software for this project is very similar to that for Chapter 24, so I will just highlight the differences. One difference is that we now have three motors to send pulses to rather than one. Each clock is driven by separate pairs of pins, which does allow for the possibility of other uses of this hardware, say, a clock that shows actual time, optimistic time, and pessimistic time. To manage the three sets of pins efficiently, they are held in an array:

```
const int pinA[] = {2, 4, 6};
const int pinB[] = {3, 5, 7};
```

The code that in Chapter 24 used individual pins now has to loop over the three sets of pins. This is illustrated in the modified setup function but also applies in the second `pulse` function.

```
void setup()
{
  for (int i = 0; i < 3; i++)
  {
    pinMode(pinA[i], OUTPUT);
    pinMode(pinB[i], OUTPUT);
  }
  t.every(second, pulse);
}
```

One final nicety that is added to the project is a constant to hold the number of milliseconds in a second. Ideally, this would be 1,000, but I found that even though the Arduino's timing is itself controlled by a quartz crystal, the clock loses a few seconds an hour. So tweaking this value allows it to keep better time.

Using the Project

The clocks can be set to the correct time zones simply by putting the hour hands on in the right places or by twiddling the knobs on the back of the mechanisms. If you find that your clock runs slow or fast, then the following procedure will allow you to find a new value for the `seconds` constant:

- Set the minute hand of the clock to 12.
- When the second hand of the clocks passes 12, start a stopwatch timing.
- After 1 hour, see how many seconds the clock has lost or gained.
- Use the following formula to find a new value for the `seconds` constant:

  Seconds = 1,000 + (seconds gained × 1,000)/3,600

In the case of my Arduino, it lost 4 seconds in an hour, so putting the numbers in the formula

Seconds = 1,000 + (–4 × 1,000)/3,600 = 998.9

the value of `seconds` must be a whole number that can be rounded to 999.

Summary

This is the last of the projects in this section. In the chapters that follow, there is a series of projects that can best be described as *novelty* projects.

PART SIX

Novelty

CHAPTER 26
Larson Scanner

Difficulty: ★★★★	Cost guide: $15

Fans of both the original and remake of the TV series *Battlestar Galactica* will be familiar with the enemy of humankind, the Cylons. These robots came in various guises, some of which had a sinister red scanning light display. The famous TV series *Knight Rider* featured a car called KIT that had a similar light display on the front. These types of LED display are called *Larson scanners* after Glen A. Larson, the producer of both the original *Battlestar Galactica* and *Knight Rider*.

This project replicates those displays using large 10-mm LEDs for maximum impact (Figure 26-1). The project is too wide to fit on a shield, so a prototyping stripboard is used.

Parts

To build this project, you will need the following:

Name	Quantity	Description	Appendix
	1	Stripboard, 8 strips by 55 holes	H5
R1–11	11	270 Ω, ¼ W resistors	R2
D1–11	11	10-mm red LEDs	S8
	1	Ribbon cable (old IDE PC cable)	
	1	Strip of 0.1-inch header sockets	H2

Figure 26-1 Larson scanner.

The ribbon cable is cut from an IDE computer cable of the sort that formerly was used to connect hard disks to the main board before Serial Advanced Technology Attachment (SATA) came along. If this is not available, then you could use separate wires, but ribbon cable of some sort helps to keep things neat.

Wiring Diagram

Figure 26-2 shows the wiring diagram for this project.

Construction

This is a theoretically simple project made harder to build because there are a lot of leads to be stripped, tinned, and connected to the stripboard. Its not as difficult as the LED cube of Chapter 2, but it requires some skill with the soldering iron to complete.

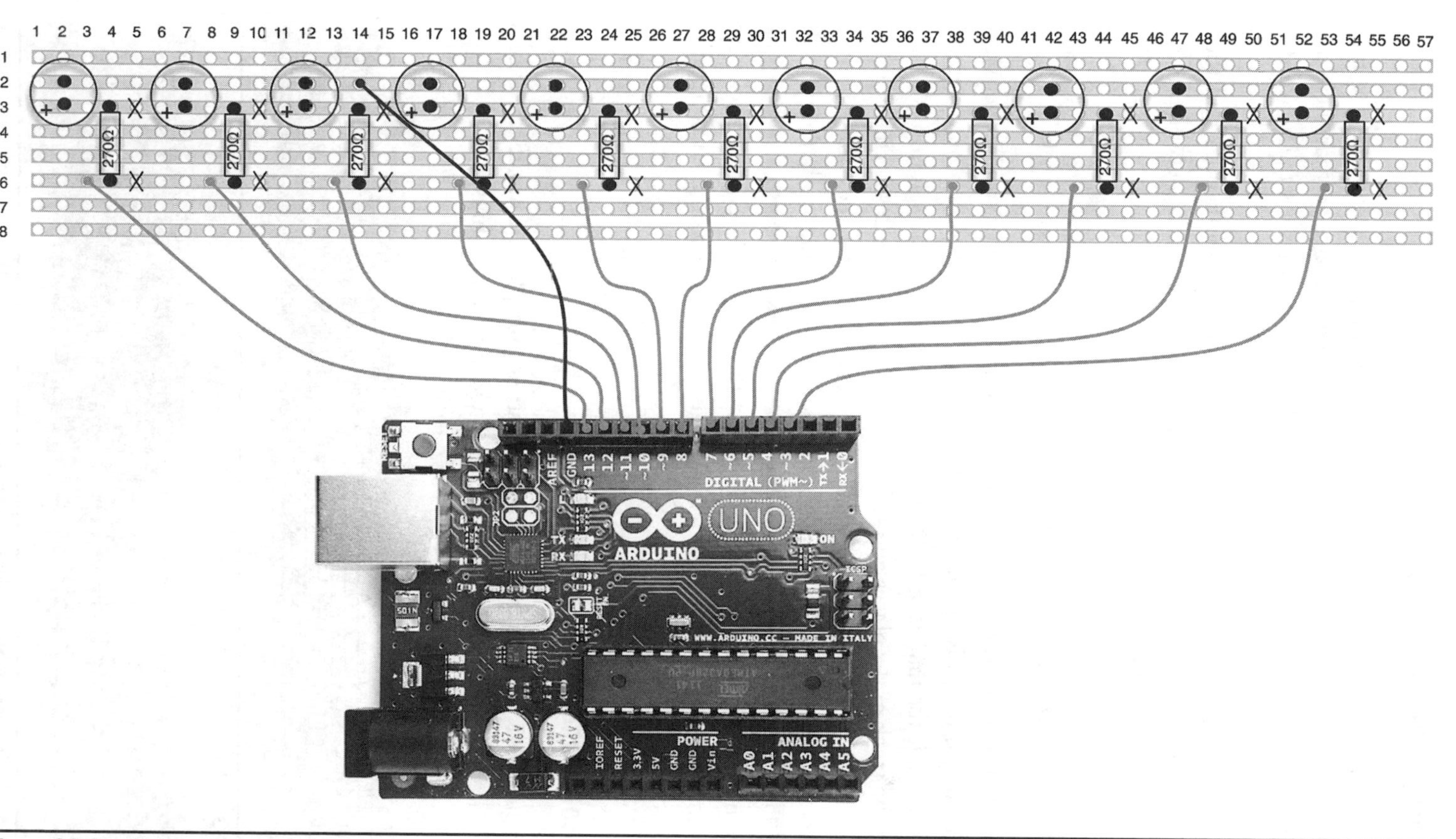

FIGURE 26-2 Wiring diagram for the Larson scanner.

Step 1: Cut the Stripboard to Size

Stripboard is a kind of ready-made PCB for prototyping. You are unlikely to find a piece just the right size, so the first step is to cut your board so that there are eight strips, each with 55 holes (Figure 26-3). It does not matter if the board is a bit larger than this.

Stripboard is not easy to cut. It is better to score it on both sides and then break it over the edge of your worktable. Do not try to break it between holes. Break it along a row of holes.

Step 2: Cut the Breaks in the Stripboard

All the LEDs are going to be in a row (and therefore connected to each other with the copper strip), which is fine for the common negative connection but no good for the positive leads that each LED needs its own resistor. The solution is to cut breaks in the track at the places marked with an "X" in Figure 26-2. This is easily done by twisting a drill bit (⅛-inch works well) between thumb and forefinger just enough to break the track. When all the breaks have been made, the board will look like Figure 26-4.

Note that Figure 26-2 shows the board from the top, but the copper tracks are underneath. So it is probably easiest to use Figure 26-4 when you are making the breaks in the copper tracks.

Step 3: Solder the Resistors

Using Figure 26-2 as a guide, solder the resistors into place. Cut off the excess leads. The board should look like Figure 26-5 when you have finished.

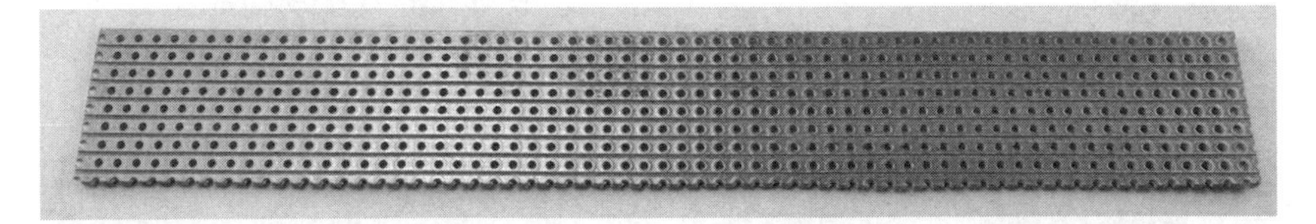

FIGURE 26-3 Stripboard cut to size.

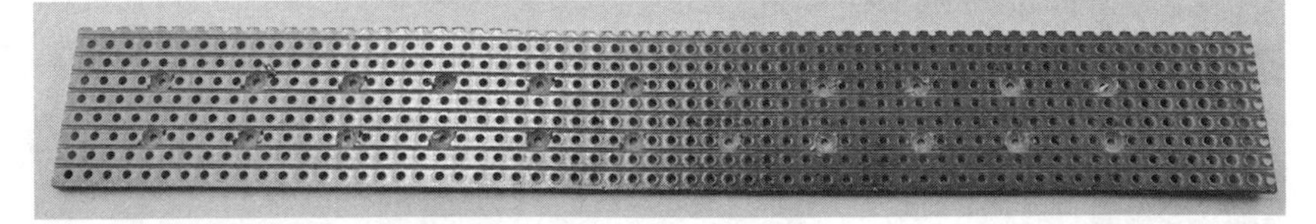

FIGURE 26-4 Stripboard with breaks in the tracks.

FIGURE 26-5 Stripboard with resistors in place.

Step 4: Solder the LEDs

The next step is to solder the LEDs. Make sure that the shorter negative leads are to the top of the stripboard (Figure 26-6).

It is best to solder just one side of each LED first. In this way, it is easier to adjust the LEDs so that they are all lying flush with the stripboard.

Step 5: Prepare the Ribbon Cable

Use scissors to cut a length of ribbon cable about 6 inches long. The individual wires will separate easily, so pull apart a ribbon of 12 wires. Splay out the ends so that they will all reach the points they need to connect to on the stripboard, and strip and tin them with solder. Do the same for the other end of the cable where it will meet the Arduino (Figure 26-7). The leads at this end do not need to be splayed out as much because the header pins are much closer together.

FIGURE 26-6 Stripboard with LEDs in place.

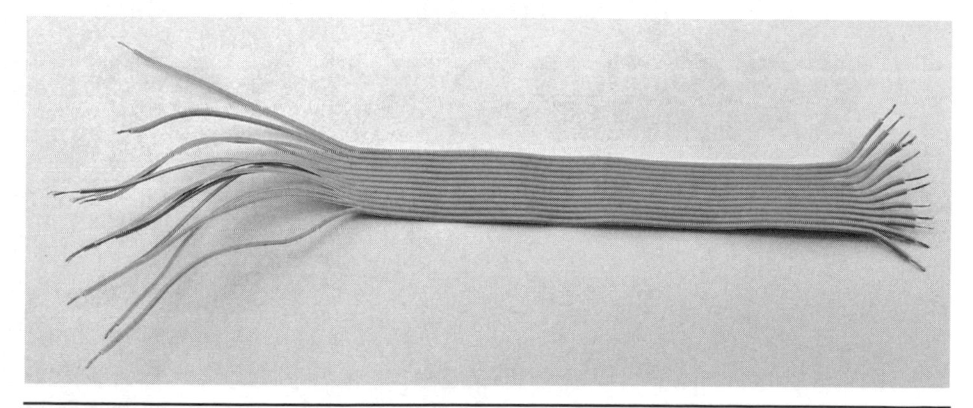

FIGURE 26-7 Ribbon cable.

Step 6: Solder the Ribbon Cable to the Stripboard

This is where it gets a bit fiddly. Solder each of the leads in turn to the correct position as shown in Figure 26-2. It helps to bend the lead over when you have poked it through the hole so that it stays in place when you turn the board over to solder it. When all the connections are made, the board should look like Figure 26-8.

Step 7: Solder the Ribbon Cable to the Header Pins

Break the header pins into one length of seven pins and one of five pins. Push them into place on your Arduino, with the 5-pin header covering pins D3–7 and the 7-pin header covering D8 to GND. Then solder the leads of the ribbon cable to the header pins. When all the soldering is complete, the board should look like Figure 26-9.

Software

The sketch for this project, ch_26_larson_scanner, is listed below. All sketches are available as a download from the author's website at www.simonmonk.org.

```
int period = 100;

void setup()
{
  for (int pin = 3; pin < 14; pin++)
```

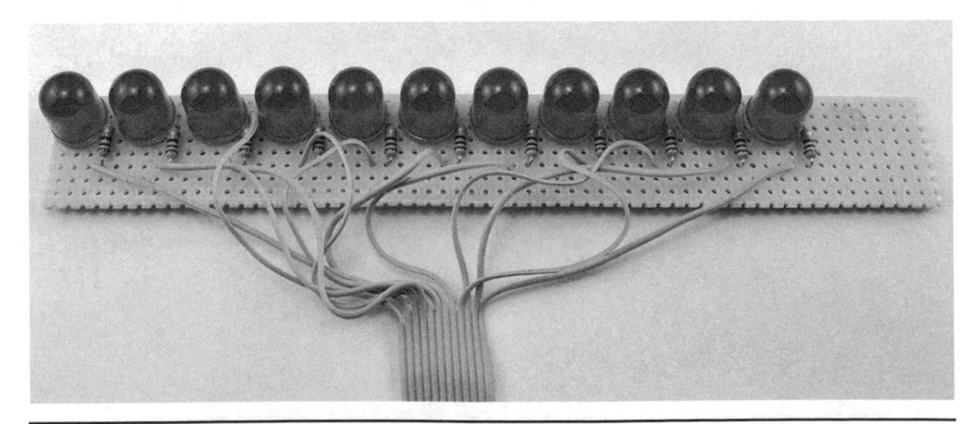

Figure 26-8 Ribbon cable attached to the stripboard.

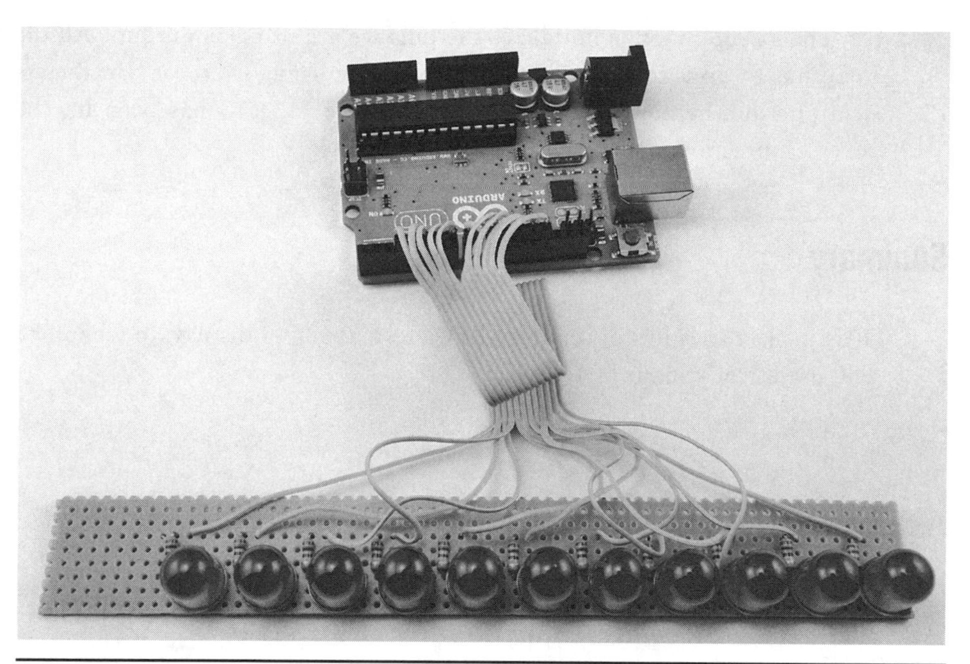

FIGURE 26-9 Project completed.

```
  {
    pinMode(pin, OUTPUT);
  }
}

void loop()
{
  for (int pin = 3; pin < 14; pin++)
  {
    digitalWrite(pin, HIGH);
    delay(period);
    digitalWrite(pin, LOW);
  }
  for (int pin = 12; pin > 3; pin--)
  {
    digitalWrite(pin, HIGH);
    delay(period);
    digitalWrite(pin, LOW);
  }
}
```

The `setup` function initializes the pins we are using as outputs. All the main `loop` has to do is to first count up each LED, turning it on for half the specified delay period, before turning it off again. Once LED 13 has been lit, the `loop` reverses direction, and so on.

Summary

This project uses lots of separate LEDs to make up a display. In Chapter 27, we will use a ready-made LCD shield.

CHAPTER 27

Conway's Game of Life

Difficulty: ★★★	Cost guide: $35

Conway's Game of Life was invented by English mathematician John Horton Conway in 1970. It is a simulation of life as a grid of cells that follow a simple set of rules as they progress through each generation (http://en.wikipedia.org/wiki/Conway's_Game_of_Life). The simulation will often reach a stable state, or all life vanish after a number of generations. In any case, it's fascinating to watch the colonies of cells evolve.

This project uses the Arduino and display to give an animated view of the game (Figure 27-1). When the joystick button is pressed or moved in any direction, the animation pauses.

Parts

To build this project, you will need the following:

Part	Quantity	Description	Appendix
	1	Adafruit 1.8-inch TFT shield	M24

The TFT display shield is supplied with header pins.

Figure 27-1 Game of Life on the Arduino.

Construction

The only hardware for this project is the TFT shield itself. The only construction required is to solder the pin headers onto the shield. Note that the shield has the old Arduino pinout rather than the latest R3 pinout. It will still fit, but some of the pins on an Arduino R3 will not be connected to the shield.

Software

This project is mostly about the software, and although the program is not long, it does use what in Arduino terms is a lot of memory. So it is a bit of a challenge to get it to work on the Arduino. The sketch is ch_27_life, and you will find it with the book downloads at www.simonmonk.org.

The sketch uses two libraries for the display—the "GFX" library that does high-level graphics and the "ST7735" library that has the low-level interfacing code with the display's chip. Download the "GFX" library from https://github.com/adafruit/Adafruit-GFX-Library and the "ST7735" library from https://github.com/adafruit/Adafruit-ST7735-Library. Once downloaded, the zip files should be extracted, copied to your Arduino "Libraries" folder, and renamed

"Adafruit_ST7735" and "Adafruit_GFX." You will need to restart the Arduino IDE for the new libraries to be picked up.

The sketch starts with the import statements for the libraries. The "SPI" library is required for communication with the display. "SPI" is a built-in library, so there is nothing to install.

```
#include <Adafruit_GFX.h>
#include <Adafruit_ST7735.h>
#include <SPI.h>
```

The display is then initialized using the following command:

```
Adafruit_ST7735 tft = Adafruit_ST7735(10, 8, -1);
//                                     TFT_CS, TFT_DC, TFT_RST
```

The constants `w` and `h` are used for the width and height of the display in cells. Although the display is 160 × 128 pixels, each cell is 8 pixels square; hence the dimensions are 20 × 16.

```
const int w = 20;
const int h = 16;
```

The constant `initialPopulation` is the number of cells to be created randomly over the screen when the simulation starts.

```
const int initialPopulation = 70;
```

The grid of cells is modeled in the three-dimension array `board`. The first dimension is used to remember both the current generation of the board and also to have a working area for the next generation. These are alternated so that at no point do we need to copy things from one array to another. The variable `g` (generation) is used to keep track of which of the two of these generations is the current generation. The other two dimensions of `board` are the columns (`y`) and rows (`x`) of the grid.

```
byte board[2][w][h];
byte g = 0;
```

I have removed the comments for the setup method that follows in the interest of brevity. The comments are from the original Adafruit example code for using the shield, and if you have problems, read the comments through because they relate to different ways of initializing different versions of the shield. The `setup` method initializes the display and clears it to black using `fillScreen`. It also changes the orientation to 3 (270 degrees of rotation) so that it is in landscape mode oriented the same way as the text labels on the shield circuit board.

```
void setup(void)
{
  tft.initR(INITR_BLACKTAB);    // initialize a ST7735S chip,
                                // black tab
  tft.fillScreen(0x0000);
  tft.setRotation(3);
  seedRandomGeneration();
}
```

The `seedRandomGeneration` function adds the number of live cells to the board contained in the `initialPopulation` constant. It also sets the seed for the random-number generator to a value read from the analog input. The analog input is not connected to anything, so it should produce a random-ish number. If you do not seed the random-number generator in this way, you will end up with the same sequence of generations every time.

```
void seedRandomGeneration()
{
  randomSeed(analogRead(A1));
  for (byte i=0; i < initialPopulation; i++)
  {
    int x = random(w);
    int y = random(h);
    board[g][x][y] = 1;
  }
}
```

The code to display the current generation iterates over every cell on the board, displaying a filled yellow rectangle if the cell is alive (indicated by a 1).

```
void displayBoard()
{
  tft.fillScreen(0x0000);
  for (int x = 0; x < w; x++)
  {
    for (int y = 0; y < h; y++)
    {
      if (board[g][x][y])
      {
       tft.fillRect(x * 8, y * 8, 8, 8, ST7735_YELLOW);
      }
    }
```

```
  }
}
```

The main loop first checks for a press of the joystick on the shield, which is connected to analog input A3. If a button is pressed, then it does nothing. Otherwise, it updates the display, calls `updateGeneration` to calculate what the next generation will look like, and then flips the generations using `g = ! g` or `g becomes not g`. Thus, if `g` is 0, it becomes 1, and vice versa.

```
void loop()
{
  if (analogRead(A3) > 512)
  {
    displayBoard();
    updateGeneration();
    g = !g;
  }
}
```

Calculating the next generation requires every cell on the board to be examined and assigned to the result of calling updateCell on that cell.

```
void updateGeneration()
{
  for (int x = 0; x < w; x++)
  {
    for (int y = 0; y < h; y++)
    {
      board[!g][x][y] = updateCell(x, y);
    }
  }
}
```

The `updateCell` function is where we find the actual rules of the game of life. This depends on how many neighbors a particular cell has (diagonals included). Thus, the rules for a cell that is currently alive are

- If the cell has fewer than two neighbors alive, then it will die.
- If it has two or three neighbors, it will stay alive.
- If it has more than three neighbors, then it will die (of overcrowding).

Dead cells will become alive if they have exactly three neighbors alive.

```
byte updateCell(int x, int y)
{
  int n = numNeighbors(x, y);
  if (board[g][x][y] == 1)
  {
    if (n < 2) return 0;
    else if (n == 2 || n == 3) return 1;
    else if (n > 3) return 0;
  }
  else
  {
    return (n == 3);
  }
}
```

Counting the neighbors is complicated by needing to cope with the edges and corners of the board, where some neighbor slots don't exist. Two nested loops are used to survey the area with a radius of one cell around the cell in question. The special cases of edges and corners, as well as the cell itself, are ignored in tallying up the number of neighbors that are alive.

```
int numNeighbors(int x, int y)
{
  int n = 0;
  for (int dx = -1; dx <= 1; dx++)
  {
    for (int dy = -1; dy <= 1; dy++)
    {
      int x1 = x + dx;
      int y1 = y + dy;
      boolean middle = (dx == 0 && dy == 0);
      if (!middle && x1 > 0 && x1 < w && y1 > 0 && y1 < h)
      {
        n += board[g][x1][y1];
      }
    }
  }
  return n;
}
```

Summary

This is one of those projects that is fun to watch in action. Possible refinements would be to add the ability to use the joystick of the shield to set up initial arrangements of cells to see what happens. You could also draw a more interesting shape than a yellow square for each cell position. In Chapter 28, we will look at taking sensor readings from a house plant to generate music.

CHAPTER 28

Singing Plant

Difficulty: ★★★	Cost guide: $20

This project is closely related to the theramin project of Chapter 11, but I decided that it belonged here in the novelty section of the book. Instead of using a range finder to alter the pitch of a musical sine wave, this project uses a technique called capacitative sensing to control the pitch (Figure 28-1).

The Arduino is attached to one leaf of a plant using a crocodile clip, and the plant (and things close to the plant) alter the pitch of the note played through the audio output socket that needs connecting to an amplifier and speakers.

Parts

To build this project, you will need the following:

Part	Quantity	Description	Appendix
	1	Protoshield printed circuit board (PCB) and header pins	A3, H1
	1	10 kΩ trimpot	R9
		10 MΩ, ¼ W resistor	R10
	1	Powered PC speakers	
	1	Strip of 0.1-inch header socket	H2
		Crocodile-clip lead[a]	
	1	3.5-mm PCB socket	H4
	1	1.5-mm drill bit	

[a] *The crocodile-clip lead has a crocodile clip on both ends.*

Figure 28-1 A musical plant.

Construction

This project uses exactly the same Protoshield layout as the project of Chapter 11. Instead of fitting an ultrasonic range finder into the header socket, a 10-MΩ resistor is pushed into the two middle socket connections.

If you are not interested in building the theramin project of Chapter 11, then you could omit the socket and wire the resistor directly onto the Protoshield. Figure 28-2 shows the Protoshield layout, assuming that you want to keep the option to make the project of Chapter 11 open, and Figure 28-3 shows the Protoshield layout if you are only interested in the singing-plant project.

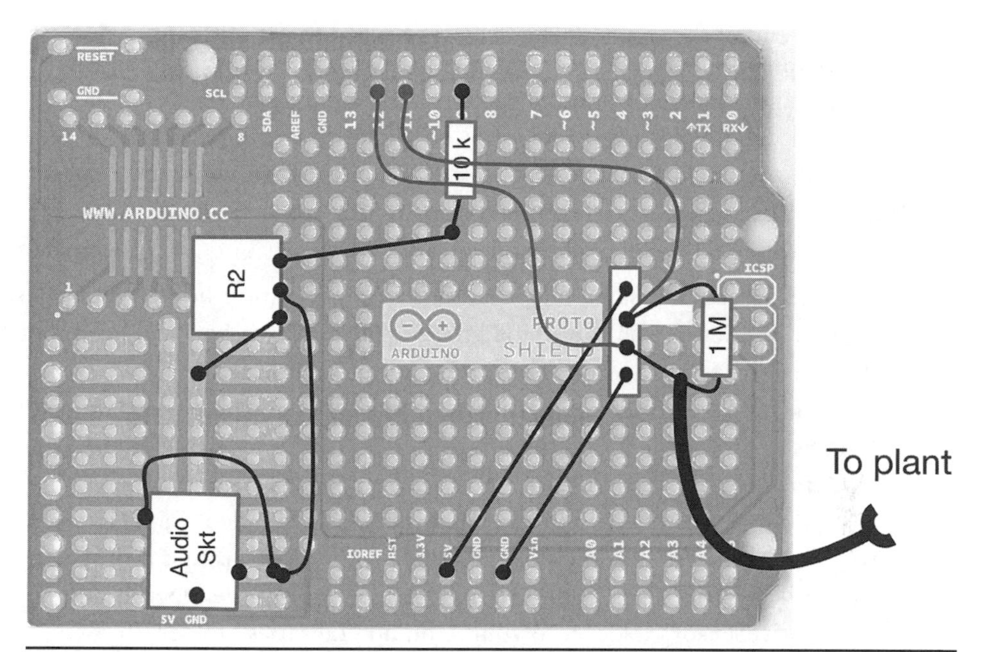

FIGURE 28-2 Protoshield layout for the singing-plant with theremin option.

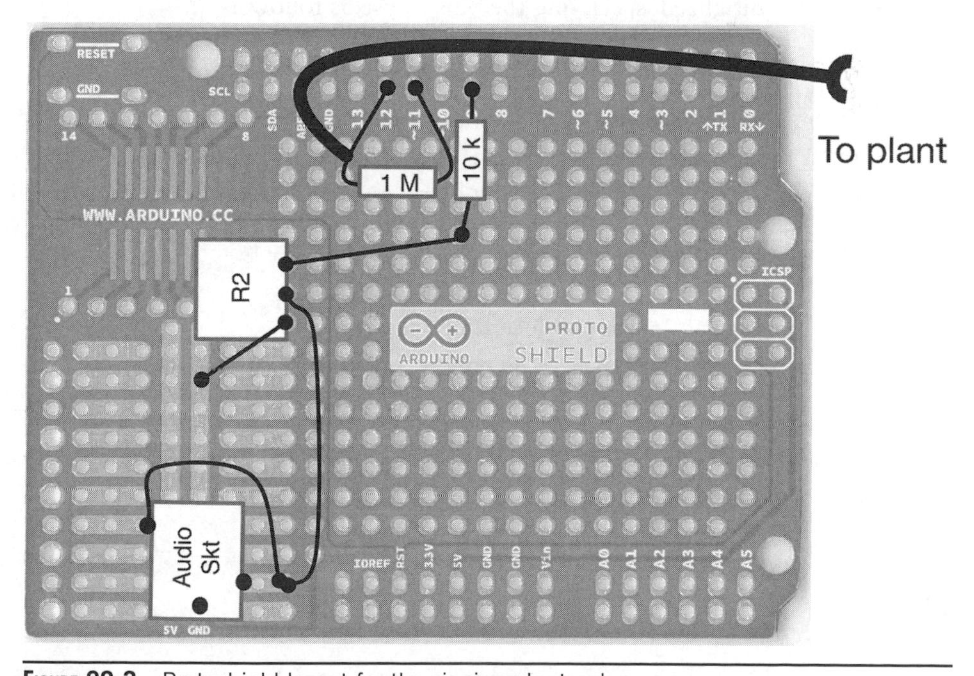

FIGURE 28-3 Protoshield layout for the singing plant only.

One end of the crocodile-clip lead is attached to the 10-MΩ resistor, as shown in Figures 28-1 through 28-3, and the other end is carefully attached to a leaf of the plant.

Software

This sketch uses the same high-quality sound-generation library that is used in Chapter 11. Before you can upload the sketch, however, you will need to get the library ("Mozzi") from http://sensorium.github.com/Mozzi/.

To install the library, select the download option that downloads the whole thing as a zip file, and save the directory into the "Libraries" folder inside your "Arduino" folder, which in turn will be in your "Documents" folder.

This project also uses a library not used in Chapter 11 called "CapacitativeSensor." You will also need to download this from http://playground.arduino.cc/Main/CapacitiveSensor. When you have downloaded the zip file, open it and copy the "CapacitiveSensor" folder into your Arduino "Libraries" folder.

The sketch itself is ch_28_singing_plant, and most of the code is very similar to that of Chapter 11. The main difference is concerned with detecting the capacitative effect of the plant using the "CapacitativeSensor" library. This library has to be initialized, specifying the pins to use as follows:

```
CapacitiveSensor cs = CapacitiveSensor(11, 12);
```

The `frequencyScaling` constant is used to adjust the raw `CapSense` reading into a range corresponding to an audio frequency. This constant may need adjusting for your plant, but for my plant, a value of 1.0 worked pretty well.

```
float takeReading()
{
  long reading = cs.capacitiveSensor(5);
  return float(reading) * frequencyScaling;
}
```

For more detail on how the audio tone generation code works, take a look at the software section of Chapter 11.

Summary

You could arrange a number of these projects to produce a whole plant orchestra! In Chapter 29, we will turn our attention to making an ultrasonic range finder.

CHAPTER 29

Ultrasonic Range Finder

Difficulty: ★★★	Cost guide: $30

This is a really quick and easy project to make. It uses an LCD shield and an HC-SR-04 range finder. The display gives a continuous readout of the distance to the nearest obstacle (Figure 29-1).

Parts

To build this project, you will need the following:

Part	Quantity	Description	Appendix
	1	Snootlabs Deuligne LCD shield	M21
	1	HC-SR-04 ultrasonic range finder	M11

The LCD shield that I used in this project is made by Snootlabs (www.snootlab.com). It is particularly useful for this project because it has through-socket headers that allow access to the pins. Most LCD shields do not offer this.

Construction

In this project, the shield is fitted onto the Arduino, and then the range-finder pins fit directly into sockets D8–11. That's it!

FIGURE 29-1 Ultrasonic range finder.

Software

The Deuligne shield does not use the standard LCD library because it has an I2C interface rather than the usual four-bit parallel interface. The functions in the library are much the same as the official library. You can download the library from https://github.com/Snootlab/Deuligne/zipball/Arduino-1.0.

The sketch (ch_29_rangefinder) starts by importing the libraries that it needs ("Wire.h" is needed for communication with the display over I2C).

```
#include "Wire.h"
#include <Deuligne.h>
```

Constants are then defined for the four pins used by the range-finder module. To make the wiring convenient, we will use pins D8 and D11 of the Arduino to supply 5 V and GND to the module. These pins are initialized in the `setup` function, which also initializes the display.

```
void setup()
{
  lcd.init();
  pinMode(trigPin, OUTPUT);
  pinMode(echoPin, INPUT);
  pinMode(supplyPin, OUTPUT);
  pinMode(gndPin, OUTPUT);
  digitalWrite(gndPin, LOW);
  digitalWrite(supplyPin, HIGH);
}
```

The `loop` function calls `takeSounding` to measure the distance and then displays it in three formats on the display.

```
void loop()
{
  lcd.clear();
  lcd.setCursor(0, 0);
  float cm = takeSounding();
  lcd.print(int(cm));
  lcd.print(" cm");
  int inches = int(cm / 2.5);

  lcd.setCursor(0, 1);
  int ft = inches / 12;
  lcd.print(ft);
  lcd.print(" ft, ");
  lcd.print(int(inches) % 12);
  lcd.print(" in");
  delay(500);
}
```

The function `lcd.setCursor` positions the cursor ready for text or other values to be displayed. Its first parameter is the column (`0` to `15`) and the second, the row (`0` or `1`). The function that actually measures the distance is as follows:

```
float takeSounding()
{
  digitalWrite(trigPin, LOW);
  delayMicroseconds(2);
  digitalWrite(trigPin, HIGH);
  delayMicroseconds(10);
  digitalWrite(trigPin, LOW);
  delayMicroseconds(2);
  long duration = pulseIn(echoPin, HIGH, 100000);
  float distance = duration / 29.0 / 2.0;
  return distance;
}
```

The `takeSounding` function sends a pulse to the TRIG pin of the range-finder module. This triggers a pulse of ultrasound. The time taken for this to return is measured, and then the distance is calculated from the time taken.

Summary

You could easily modify the format and units of the data displayed. You will also find this range finder used back in Chapter 11 to make a musical instrument. This display is also used in Chapter 30, where it is used to display GPS data.

CHAPTER 30
GPS

Difficulty: ★	Cost guide: $65

This is another quick and easy project to make. It uses the same LCD shield as Chapter 29 (range finder) and a GPS module from Adafruit. The project gives a simple readout of the latitude and longitude (Figure 30-1).

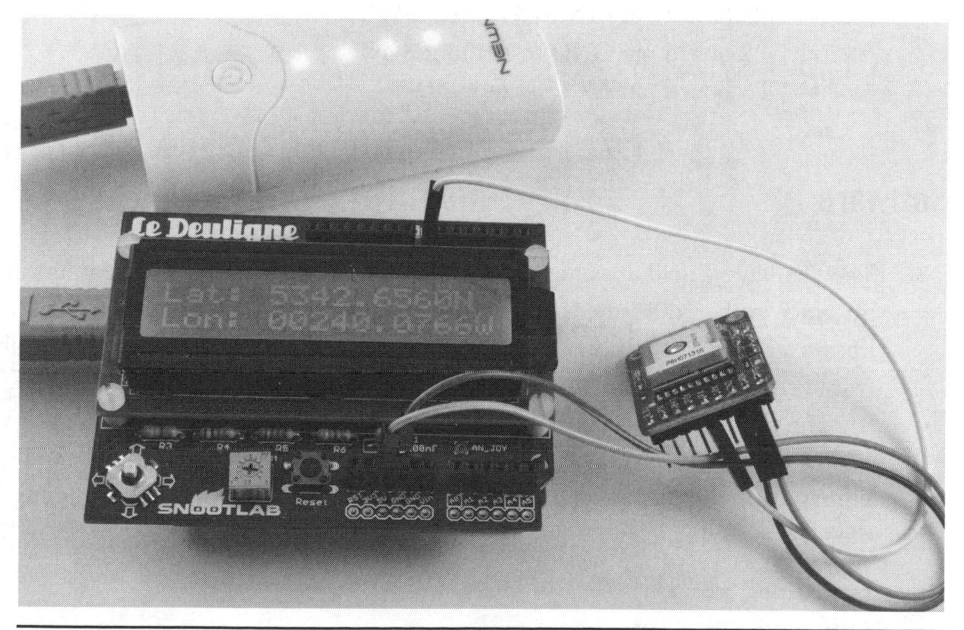

FIGURE 30-1 GPS.

Parts

This project needs an Arduino Uno because the display shield will not work with a Leonardo. To build this project, you will need the following:

Part	Quantity	Description	Appendix
	1	Snootlabs Deuligne LCD shield	M21
	1	Adafruit GPS module	M22
	3	Male-to-female jumper leads	H6

The LCD shield that I used in this project is made by Snootlabs (www.snootlab.com). It is particularly useful for this project because it has through-socket headers that allow access to the pins. Most LCD shields do not offer this.

Construction

In this project, the shield is fitted onto the Arduino, and then the GPS module is connected to the shield using three male-to-female jumper leads. The wiring is as follows:

- GPS board pin VIN to Arduino 5 V
- GPS board pin GND to Arduino GND
- GPS board pin TX to Arduino D7

Software

The Deuligne shield does not use the standard LCD library because it has an I2C interface rather than the usual four-bit parallel interface. The functions in the library are much the same as the official library. You can download the library from https://github.com/Snootlab/Deuligne/zipball/Arduino-1.0.

The sketch for this project is in ch_30_gps. It starts by importing the "Wire" and "Deuligne" libraries needed by the display shield. It also imports the built-in library "SoftwareSerial" that is used to receive the serial messages coming from the GPS module.

```
#include "Wire.h"
#include <Deuligne.h>
#include <SoftwareSerial.h>
```

Constants are defined for the pins used as well as a string buffer called `sentence`, in which the messages from the GPS module will be stored prior to picking out the latitude and longitude from them. Note that although a pin is defined for `tx` (transmit), it is not connected to anything because the Arduino only needs to receive messages transmitted from the GPS module. Communication in the opposite direction is not needed.

```
const int rxPin = 7;
const int txPin = 8; // not used
const int sentenceSize = 80;
char sentence[sentenceSize];
```

The `setup` function both initializes the display and starts the Arduino listening for messages from the GPS unit.

```
void setup()
{
  lcd.init();
  gps.begin(9600);
  lcd.clear();
  lcd.print("Starting..");
}
```

The main loop listens for characters coming from the GPS module and appends them to the `sentence` buffer.

The messages conform to a standard from the National Marine Electronics Association (NMEA). Each message is a string of text ending with the newline character. The fields of the message are separated by commas. A typical message is as follows:

```
$GPRMC,081019.548,A,5342.6316,N,00239.8728,W,000.0,079.7,
110613,,,A*76
```

You can find a compete list of the NMEA GPS sentences listed at http://aprs.gids.nl/nmea/. The fields in the preceding example are

- `$GPRMC`—the sentence type
- `081019.548`—the time (very accurate) 8:10:19.548
- `5342.6316`—North latitude × 100, that is, 53.426316 degrees North
- `00239.8728`—West longitude × 100, that is, 0.2398728 degrees West
- `000.0`—speed
- `079.7`—course 79.7 degrees

- 110613—date, that is, June 11, 2013
- The remaining fields are not of interest.

The sentences coming from the GPS module are of differing lengths but are all less than 80 characters, so the code uses a buffer variable sentence that is filled with the data until an end-of-line marker is read or the buffer is full. A null character is placed on the end of the buffer when the whole sentence has been read to mark the end of the text.

```
void loop()
{
  static int i = 0;
  if (gps.available())
  {
    char ch = gps.read();
    if (ch != '\n' && i < sentenceSize)
    {
      sentence[i] = ch;
      i++;
    }
    else
    {
     sentence[i] = '\0';
     i = 0;
     displayGPS();
    }
  }
}
```

The rest of the sketch is concerned with extracting individual fields and formatting the output to be written to the serial monitor. The getField function helpfully extracts the text from a field at a particular index.

```
void getField(char* buffer, int index)
{
  int sentencePos = 0;
  int fieldPos = 0;
  int commaCount = 0;
  while (sentencePos < sentenceSize)
  {
    if (sentence[sentencePos] == ',')
```

```
    {
      commaCount ++;
      sentencePos ++;
    }
    if (commaCount == index)
    {
      buffer[fieldPos] = sentence[sentencePos];
      fieldPos ++;
    }
    sentencePos ++;
  }
  buffer[fieldPos] = '\0';
}
```

The function `displayGPS` formats the various fields from the sentence, displaying them on the two lines of the LCD display.

```
void displayGPS()
{
  Serial.println(sentence);
  char field[80];
  getField(field, 0);
  if (strcmp(field, "$GPRMC") == 0)
  {
    lcd.setCursor(0, 0);
    lcd.print("Lat: ");
    getField(field, 3);  // number
    lcd.print(field);
    getField(field, 4); // N/S
    lcd.print(field);

    lcd.setCursor(0, 1);
    lcd.print("Lon: ");
    getField(field, 5);  // number
    lcd.print(field);
    getField(field, 6);  // E/W
    lcd.print(field);
  }
}
```

Note that the GPS module will spit out various sorts of sentence. Each one starts with an identifier, so when displaying the message, all but messages starting with `$GPRMC` are ignored.

Using the Project

You will need to take the module outside or at least place it right by a window so that it can acquire a satellite fix. To make it portable, a USB battery booster designed for mobile phones makes a useful rechargeable power source. You can also use a 9-V battery with a battery clip to barrel jack adapter.

Summary

This is a project that could be modified, for example, to display the time from the GPS satellite, which is extremely accurate. In Chapter 31, we will look at detecting noxious gases.

CHAPTER 31
Methane Detector

Difficulty: ★★★	Cost guide: $20

This project is one that lends itself to the childish application of detecting farts. It uses a low-cost methane sensor to detect the presence of this gas, and it sounds a buzzer and flashes a bright LED when it is detected (Figure 31-1).

FIGURE 31-1 Methane gas detector.

Parts

To build this project, you will need the following:

Part	Quantity	Description	Appendix
	1	Protoshield and header pins	A3, H1
	1	MQ4 methane sensor	M20
	1	Piezo buzzer	C3
	1	10-mm red LED	S8
R1		20 kΩ, ¼ W resistor	R11
R2		270 Ω, ¼ W resistor	R1
R3		10 kΩ trimpot	R9

Construction

In this project, all the components are mounted onto a Protoshield. Figure 31-2 shows the layout of the components.

The methane sensor is a device whose resistance changes in the presence of methane. Figure 31-3 shows how one of these devices is arranged internally.

The sensor has a catalytic strip that senses the gas and a heater that warms the strip up. The heater requires 5 V and will draw a maximum of about 150 mA, which is just fine for the Arduino to supply from the 5-V rail. Because both the sensing element and the heater are essentially just resistors, the sensor can be fitted either way around.

Step 1: Solder the Pins to the Protoshield

This project does not use the row of pin headers from D0 to D7, so these do not need to be attached to the Protoshield.

Step 2: Attach Leads to the Methane Sensor

The methane sensor module has rather fat little legs that will not fit through the holes on a Protosheild, so solder short lengths of solid-core wire to them, as shown in Figure 31-4.

Step 3: Solder the Resistors

Start by soldering the two resistors to the board (Figure 31-5). Do not trim off the resistor leads because they will be needed for making connections on the underside of the board.

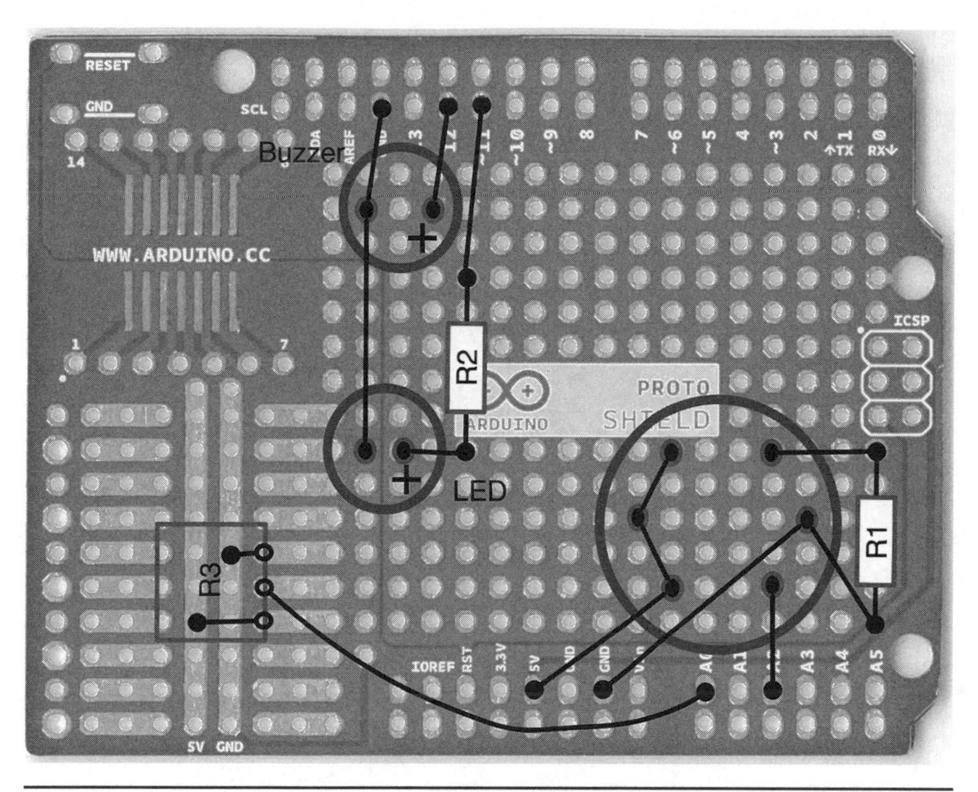

FIGURE 31-2 Protoshield layout for the methane sensor.

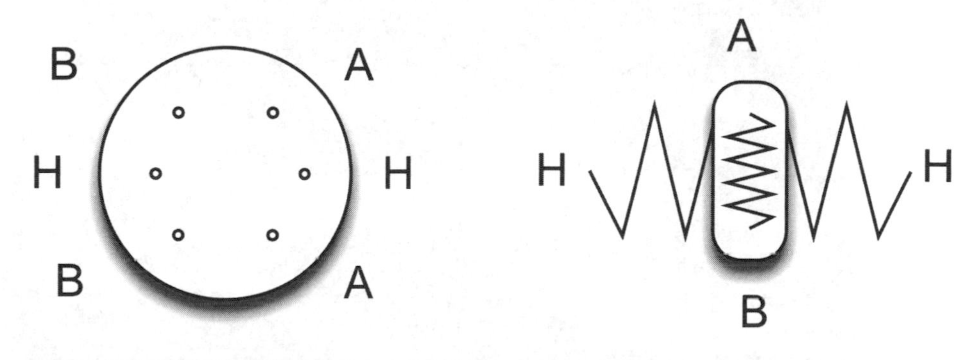

FIGURE 31-3 MQ4 methane sensor.

Figure 31-4 Attaching leads to the methane sensor.

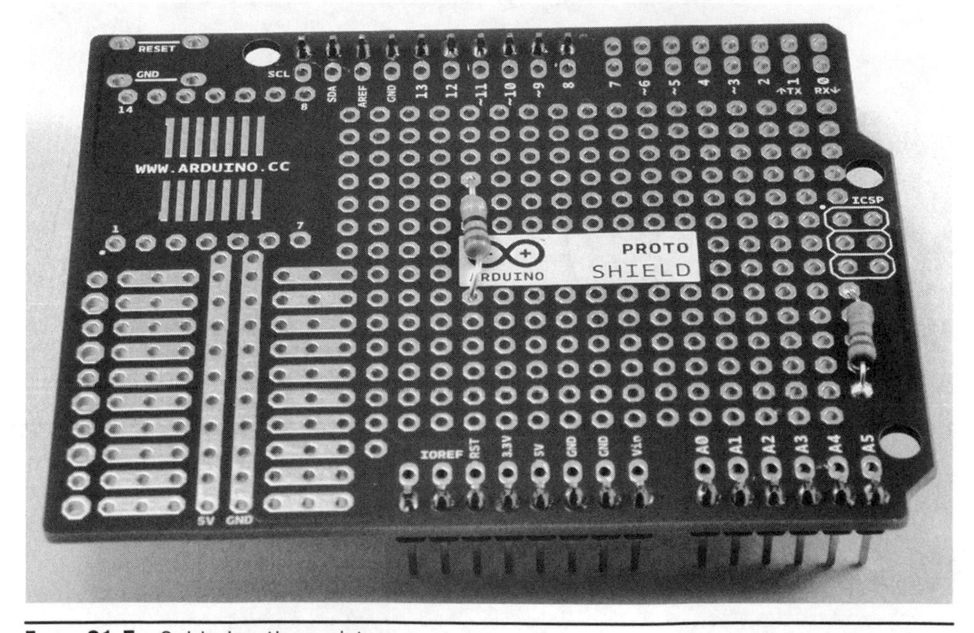

Figure 31-5 Soldering the resistors.

Step 4: Solder the Remaining Components

Now solder the rest of the components onto the shield, starting with the lowest component—the buzzer. The buzzer may have one pin marked with a minus sign ("–"). Make sure that is in the same position as shown in Figure 31-2. If there is no marking, then the longer lead will be positive ("+"). If there is no marking and neither lead is longer than the other, then it does not matter which way around you solder the buzzer.

The longer lead of the LED should go to positive, as shown in Figure 31-2. When all the components are in place, the top of the Protoshield will look like Figure 31-6.

Step 5:. Wire the Underside

Figure 31-7 shows how the board is wired up from the underside.

Wherever possible, bend the longer leads from the resistors and LED to make connections. Where these will not reach, you will have to cut short lengths of solid-core wire. Use insulated wire when these have to cross, and pay special attention around the trimpot because a short length of wire is needed to connect one side of the trimpot to 5 V, and this must jump over a ground line without connecting to it. When all the connections are made, the bottom of the board should look like Figure 31-8.

FIGURE 31-6 Protoshield with all the components attached.

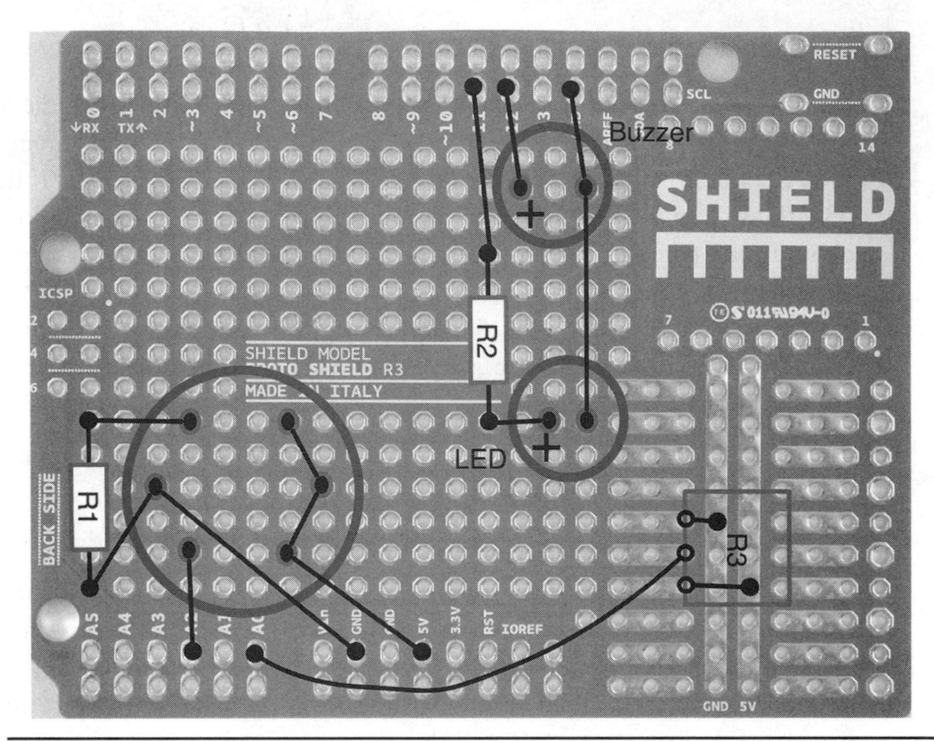

Figure 31-7 Wiring diagram for the underside of the board.

Figure 31-8 Underside of the Protoshield.

Software

The software for this project can be found in the sketch ch_31_fart_detector. Constants are used to define the pins for the sensor, the trimpot, the buzzer, and the LED.

```
const int sensorPin = A2;
const int trimpotPin = A0;
const int buzzerPin = 12;
const int ledPin = 11;
```

The `setup` function sets the pin modes and then initializes serial communication. Serial communication is only used as a debugging tool, displaying the readings from the sensor.

```
void setup()
{
  pinMode(buzzerPin, OUTPUT);
  pinMode(ledPin, OUTPUT);
  Serial.begin(9600);
  while(! Serial) {};
  Serial.println("Warming Up...");
  warmUp();
  Serial.println("Ready");
}
```

The `wamUp` function called by `setup` allows the heater to warm up the sensor to its working temperature. This takes a few minutes depending on the ambient temperature. During this period, the sensor readings will gradually fall. The function detects when the readings stop falling and actually rise a tiny bit at the end of the warming process. During the warming process, this function also makes the LED blink.

```
void warmUp()
{
  int d = -1;
  int oldReading = 1023;
  while (d <= 0)
  {
    int reading = analogRead(sensorPin);
    Serial.println(reading);
```

```
      d = reading - oldReading;
      oldReading = reading;
      digitalWrite(ledPin, HIGH);
      delay(200);
      digitalWrite(ledPin, LOW);
      delay(200);
    }
  }
```

The loop function takes readings of the trimpot and the sensor, and if the sensor reading is greater than the value set by the trimpot, the function causes the LED to light and the buzzer to sound using the Arduino tone function.

```
void loop()
{
  int reading = analogRead(sensorPin);
  int threshold = analogRead(trimpotPin);
  if (reading > threshold)
  {
    tone(buzzerPin, 500);
    digitalWrite(ledPin, HIGH);
  }
  else
  {
    noTone(buzzerPin);
    digitalWrite(ledPin, LOW);
  }
}
```

Using the Project

You do not need actual methane to test this project; breathing on it will have enough of an effect. Power up the Arduino, and when warming has finished and the LED stops flashing, adjust the trimpot until the LED and buzzer are just off. Now, when you breathes gently on the sensor, you should trigger the LED and buzzer.

Summary

The methane sensor readings need to be moderated with readings of ambient humidity and temperature to be accurate, but nonetheless, this is a fun little project. In Chapter 32, we will change the mood somewhat, beginning a new section on health and well-being.

PART SEVEN

Home

CHAPTER 32

Light-Level Logger

Difficulty: ★★	Cost guide: $25

The goal of this project is to build a simple tool for logging light intensity against time. The idea is that when planning your garden planting, you can have a reasonable idea of how much sunlight an area receives. Ideally, you would want to compare readings in different locations over the same 17-hour period, so you would need more than one of the devices.

The figure 17 hours is not completely arbitrary. It will cover the daylight hours in most parts of the world, and you will be able to store one reading per minute into the electrically erasable programmable read-only memory (EEPROM) for that period without having to resort to external storage. Figure 32-1 shows one of the devices unboxed.

Parts

To build this project, you will need the following:

Part	Quantity	Description	Appendix
	1	Protoshield and header pins	A3, H1
	1	1 kΩ photoresistor	R6
R1	1	1 kΩ resistor	R2
R2	1	270 Ω resistor	R1

(continued on next page)

	1	Red LED	S8
S1	1	Push switch	C2
	1	USB backup battery pack	
	1	Table tennis ball or diffuser (optional)	

Photoresistors vary considerably in resistance range. The one that I used had a "light" resistance of 1 kΩ. If you have a photoresistor of higher resistance, pick a resistor of the same value for R1. The LED can be any color or size; it is not critical.

In practical use, you can add half a table tennis ball or some other hemispherical translucent diffuser over the top of the photoresistor. This is by no means essential,

FIGURE 32-1 Light logger.

and you can just place the whole unit in a translucent plastic food storage box, which will have the added advantage of protecting it from the elements.

The USB battery pack is of the sort used to provide backup power to a cell phone. The idea is that you connect it to the USB socket of your computer or USB charger, and it charges up its internal battery, and you can then connect a USB lead to it and charge or run your phone (or an Arduino). Be warned, though, some of these devices will be fooled by the low power consumption of the Arduino and turn themselves off after a while.

Construction

Figure 32-2 shows the Protoshield layout for the project.

Step 1: Attach the Header Pins to the Protoshield

This project uses all the pin headers except the top right D0–7, which you can leave off if you prefer.

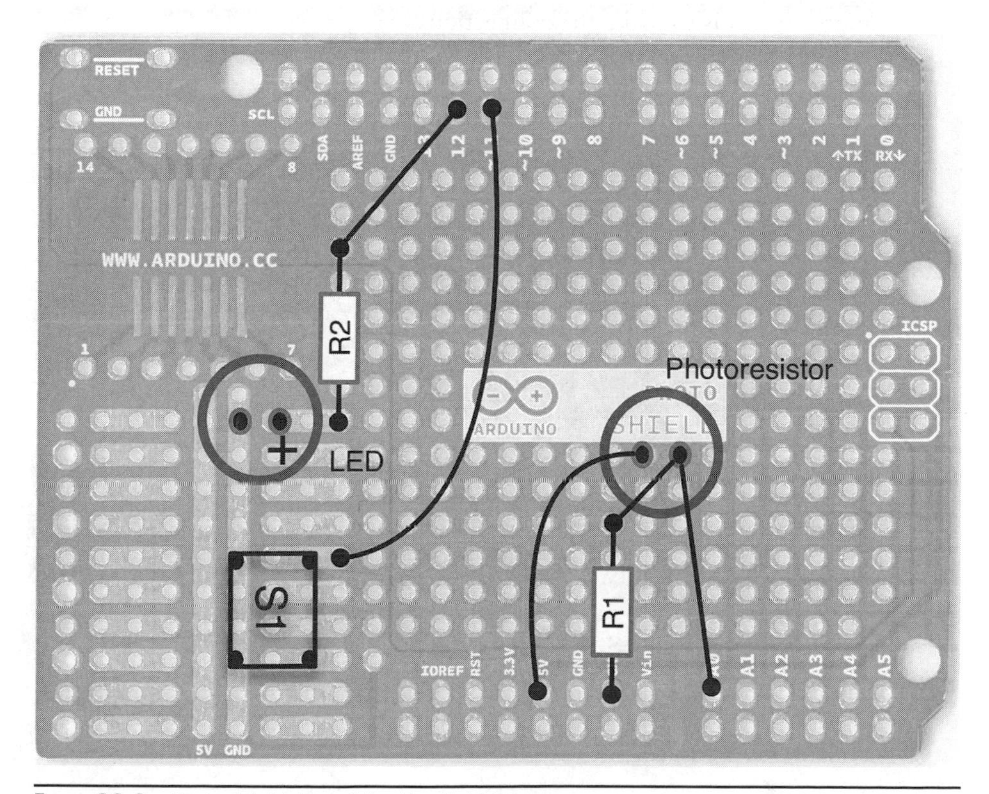

Figure 32-2 Protoshield layout for the light logger.

Step 2: Solder the Resistors to the Protoshield

Solder R1, R2, and the phototresistor to the Protoshield. Do not snip off the leads on the underside of the board because these can be used to connect things up. Note that it does not matter which way around the photoresistor is placed. When this is done, the board should look like Figure 32-3.

Step 3: Solder the Remaining Components to the Protoshield

You can now solder on the switch and the LED. The switch will only fit one way around in the holes indicated in Figure 32-2. The LED should have its shorter negative lead to the left connected to the GND line on the Protoshield. With these components in place, the board should look like Figure 32-4.

Step 4: Solder the Underside of the Protoshield

Using Figure 32-5 as a reference, solder the underside of the board. Apart from one longer wire to connect the switch to D11, you should be able to use the component leads to make the connections.

Figure 32-3 Resistors soldered in place.

FIGURE 32-4 Board with all components in place.

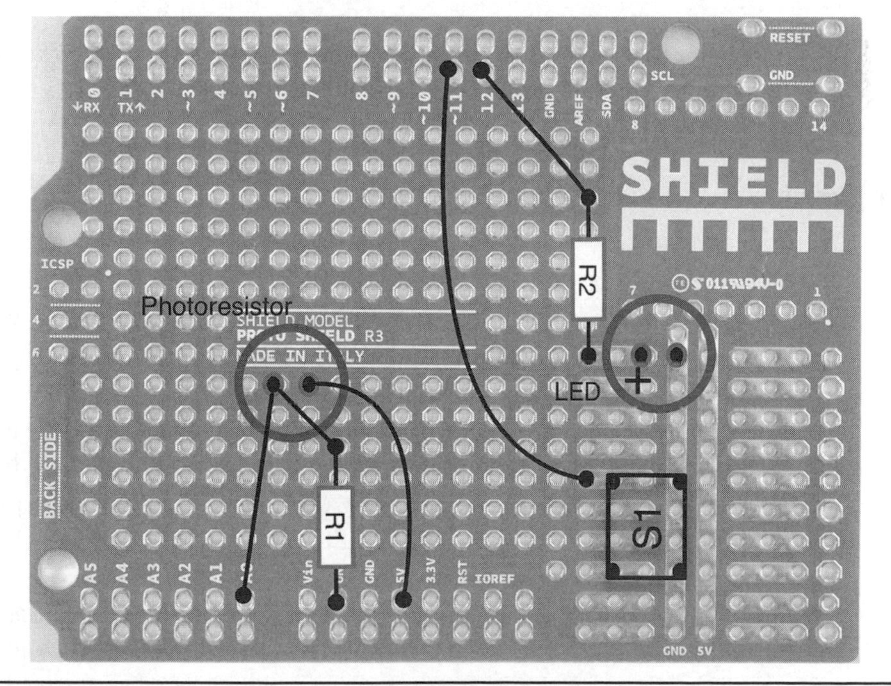

FIGURE 32-5 Wiring diagram for the underside of the board.

Figure 32-6 Underside of the Protoshield.

Remember to cut off the excess leads when everything is wired. When this has been done, the underside of the board should look like Figure 32-6.

Software

The sketch for this project is ch_32_light_logger, and it uses the built-in "EEPROM" library to store the readings in the EEPROM so that they are not lost when the Arduino resets. The sketch also uses the "Timer" library, which can be downloaded from https://github'com/JChristensen/Timer.

When you have downloaded the zip file, you will have to rename the unzipped file as just "Timer" and copy it into your Arduino "Libraries" folder. After importing the libraries, the sketch starts by defining constants for the pins used and the sample period in seconds. This is set to 60 seconds, which will allow over 17 hours of recording. Note that the Arduino has just 1,024 bytes of EEPROM in which you can sore readings, so you cannot increase the value of `maxNumReadings` to take more readings. If you want to record for a longer period, you will need to increase the `readingPeriod` constant, so that readings are taken less often.

```
const int sensorPin = A0;
const int ledPin = 12;
const int maxNumReadings = 1024;
const int switchPin = 11;
const long readingPeriod = 60; // 60 seconds (17 hours worth)
```

A Boolean variable `recording` is used to keep track of whether recording is in progress, and the variable `readingIndex` keeps track of where to store the next reading. The variable `t` is used for the timer, which will control both the taking of readings and the blinking of the LED.

```
boolean recording = false;
int readingIndex = 0;
Timer t;
```

The `setup` function sets the pin modes and starts serial communication, which is used when getting the logged data off the Arduino. It also sets two functions to be run at intervals. The first function to be run every `readingPeriod` is `takeReading`. The second is `updateLED`, which will be run every second.

```
void setup()
{
  pinMode(ledPin, OUTPUT);
  pinMode(switchPin, INPUT_PULLUP);
  Serial.begin(9600);
  while (! Serial) {};
  Serial.println("d-download c-clear");
  t.every(readingPeriod * 1000L, takeReading);
  t.every(1000, updateLED);
}
```

The function `takeReading` will (if `recording` is `TRUE`) read a value from the analog input and then process it for storage as a single byte in the EEPROM.

```
void takeReading()
{
  if (recording && readingIndex < maxNumReadings)
  {
    digitalWrite(ledPin, LOW);
    int raw = analogRead(sensorPin);
    // 0 to 1024 log10 0 to 3
    byte reading = byte(log10(float(raw)) * 80.0);
```

```
      Serial.println(reading);
      EEPROM.write(readingIndex, reading);
      readingIndex++;
    }
}
```

The photoresistor measures light in a logarithmic manner, so to give us a better range of values, we take the log to the base 10 and scale it so that it will always fit into a number less than 256. This means that we can then store it as a single byte using the `EEPROM.write` function. The `updateLED` function makes the LED give a very brief blink once per second when the device is recording to provide feedback that everything is working.

The `loop` function checks for commands arriving from the serial monitor, calling the appropriate functions `dump` and `clear`. It also checks for the button being pressed that will start recording. The call to `t.update()` allows the timer to check whether anything needs doing.

```
void loop()
{
  if (Serial.available())
  {
    char ch = Serial.read();
    if (ch == 'd')
    {
      recording = false;
      dump();
    }
    else if (ch == 'c')
    {
      recording = false;
      clear();
    }
  }
  if (digitalRead(switchPin) == LOW)
  {
    recording = true;
  }
  t.update();
}
```

The `clear` function simple zeros in every slot in EEPROM storage, and the dump function sends all the stored values to the serial monitor, from where they can be copied and pasted into a spreadsheet (see the next section).

```
void clear()
{
  for (int i = 0; i < maxNumReadings; i++)
  {
    EEPROM.write(i, 0);
  }
  readingIndex = 0;
  Serial.println("Cleared");
}

void dump()
{
  for (int i = 0; i < maxNumReadings; i++)
  {
    Serial.println(EEPROM.read(i));
  }
}
```

Using the Project

To use this project for the first time, you can place it with the USB power supply in a weatherproof translucent box. Turn on the power source, and press the button on the Protoshield. Logging should then start, indicated by the LED flashing. A 4-Ah battery unit should be able to power an Arduino for up to 100 hours.

When you are ready to take the readings from the Arduino, connect it to your computer and start the serial monitor from the Arduino IDE (Figure 32-7).

This prompts you to enter one of the single-letter commands `d` or `c`. To print out all the readings to the serial monitor, enter the `d` command. Scroll back up to the top of the readings, select the range of readings that you want to use, and then paste them into a spreadsheet where you can process or chart them however you like (Figure 32-8).

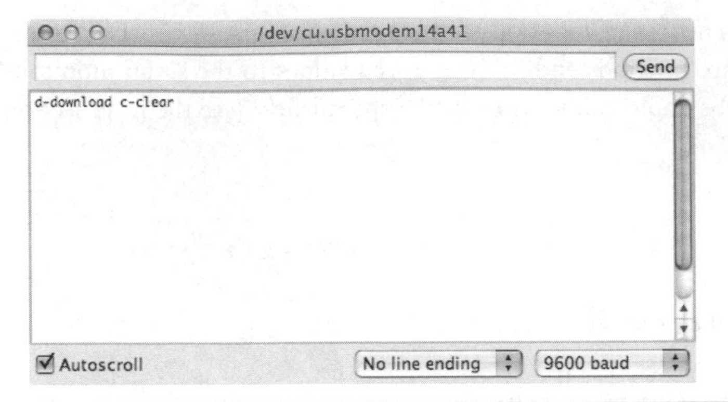

FIGURE 32-7 Commands from the serial monitor.

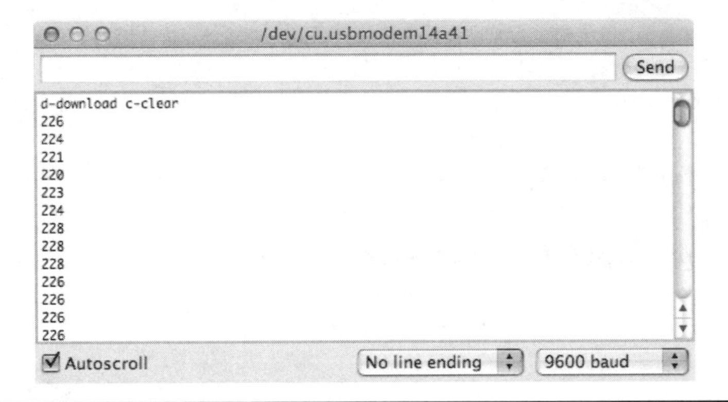

FIGURE 32-8 Light readings used in a spreadsheet.

Summary

In Chapter 33, we will extend this project to add temperature sensing as well as light sensing.

CHAPTER 33

Temperature and Light Logger

Difficulty: ★★	Cost guide: $25

This project builds on the project described in Chapter 32, adding the ability to log temperature at the same time as it logs the light level (Figure 33-1). If you only want to log temperature, then simply leave the photoresistor and R1 off the design.

Parts

To build this project, you will need the following:

Part	Quantity	Description	Appendix
	1	Protoshield and header pins	A3, H1
	1	1 kΩ photoresistor	R6
IC1		TMP36 temperature sensor	S13
R1	1	1 kΩ, ¼ W resistor	R2
R2	1	270 Ω resistor	R1
	1	Red light-emitting diode (LED)	S8
S1	1	Push switch	C2
	1	USB backup battery pack	
	1	Table tennis ball or diffuser (optional)	

FIGURE 33-1 Completed temperature and light logger.

Construction

The construction starts exactly the same as in Chapter 32. So start by following the steps described there, ignoring the area around the photoresistor and R1 if you only want to log temperature. Figure 33-2 shows the Protoshield layout for the project.

The three pins of the temperature sensor are connected to three of the analog pins, two of which are used to supply power to the chip. Make sure that you get the chip the right way around, with its curved edge toward the outside of the Protoshield.

Software

The software is mostly identical to that of the light-only logger from Chapter 32, so here I will only highlight the differences. The sketch is ch_33_temp_logger and uses the same libraries as the preceding project.

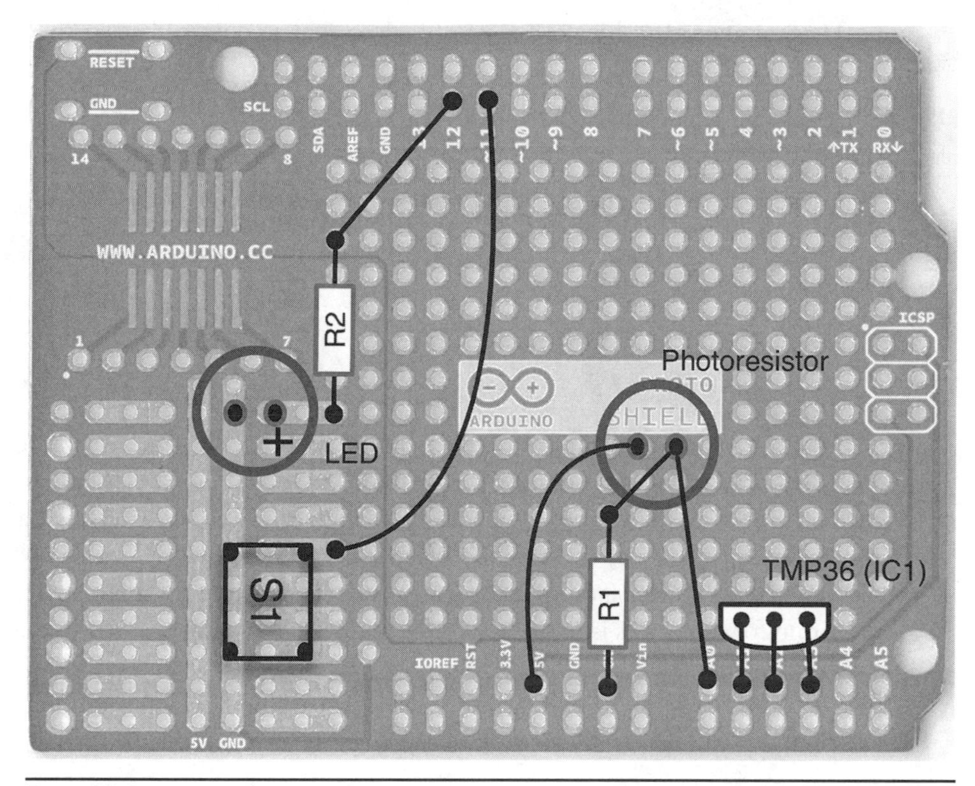

FIGURE 33-2 Protoshield layout.

Three new constants are needed for the temperature sensor pins, and the pin used for the light sensor is renamed from `sensorPin` to `lightSensorPin`.

```
const int lightSensorPin = A0;
const int tempSensorPin = A2;
const int tempGndPin = A1;
const int tempPlusPin = A3;
```

The sample period also has been altered to 120 seconds, and because we are recording twice as much data, we can only record samples at half the frequency.

The `setup` function is extended to set up the pins for the temperature sensor, with the following lines added to it:

```
  pinMode(tempGndPin, OUTPUT);
  pinMode(tempPlusPin, OUTPUT);
  digitalWrite(tempPlusPin, HIGH);
```

The `takeReading` function is modified so that it now calls two functions, `takeLightReading` and `takeTempReading`.

```
void takeReading()
{
  if (recording && readingIndex < maxNumReadings - 1)
  {
    takeLightReading();
    takeTempReading();
  }
}
```

The code for `takeLightReading` is more or less the same as that in the preceding project, except that now we are taking two readings rather than one, so we will alternate light and temperature readings in the electrically erasable programmable read-only memory (EEPROM).

The temperature readings are compressed to fit into a single byte. The algorithm for doing this is to add 20 to the temperature in degrees centigrade and then divide the result by 4. This allows a range of temperatures from –20 to +43.75°C with a resolution of ¼ degree. This is just fine for most situations, and the TMP36 temperature sensor is not very accurate anyway.

```
void takeTempReading()
{
    float tempC = readTemp();
    byte reading = byte((tempC + 20.0) * 4);
    Serial.println(tempC);
    EEPROM.write(readingIndex, reading);
    readingIndex++;
}
```

The temperature reading from the TMP36 chip is converted from a voltage to a temperature in degrees centigrade in the function `readTemp`.

```
float readTemp()
{
  int a = analogRead(tempSensorPin);
  float volts = a / 205.0;
  float temp = (volts - 0.5) * 100;
  return temp;
}
```

Now, when the data are "dumped" for import to a spreadsheet, it will be arranged in rows, with first a light level and then a temperature reading separated by a comma character. This format is called *comma-separated values* (CSVs), and most spreadsheet software will import this directly.

The temperature has to be uncompressed from the reading in the EEPROM to give a temperature in degrees centigrade using the function `getTempReading`.

```
float getTempReading(int index)
{
  byte compressedReading = EEPROM.read(index);
  float uncompressesReading = (compressedReading / 4.0) - 20.0;
  return uncompressesReading;
}
```

Using the Project

When you are ready to start recording with the device, connect it to your computer, and issue the command `c` through the serial monitor. You can then unplug the device and attach it to the USB battery pack. Then press the button to start the recording.

When you are ready to take the readings from the device, unplug it from the power source, plug it into your computer, and send the command `d` to download the data into the serial monitor.

```
208,14.25
211,14.75
205,13.25
216,12.75
207,13.25
```

When you paste the data into the spreadsheet, you may be offered the services of an "import wizard" such as Figure 33-3.

Specify that you want to separate the fields with commas, and import the data. You can now chart the data in whatever way you want within the spreadsheet. An example is given in Figure 33-4.

This manipulation might involve adding a new column to the data where the temperatures in degrees centigrade are converted into degrees Fahrenheit using the formula $C \times 9/5 + 32$.

If the photoresistor is not soldered onto the Protoshield because you were only interested in the temperature, then the light column will be garbage, and you can

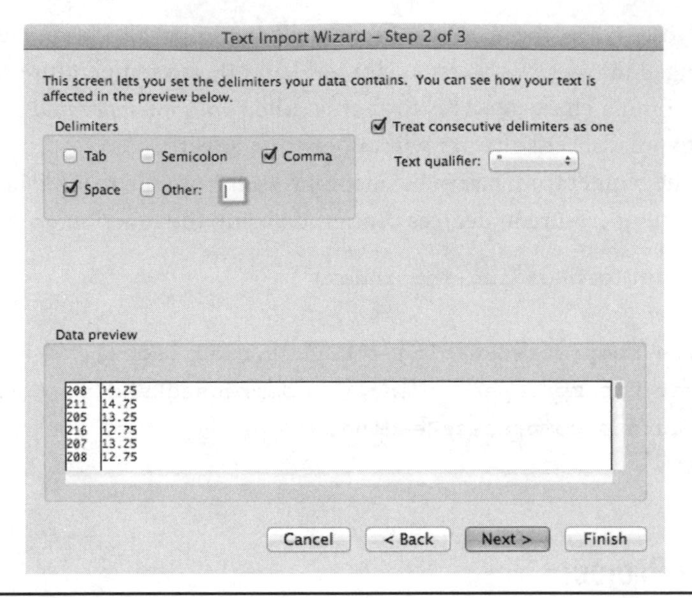

Figure 33-3 Import wizard.

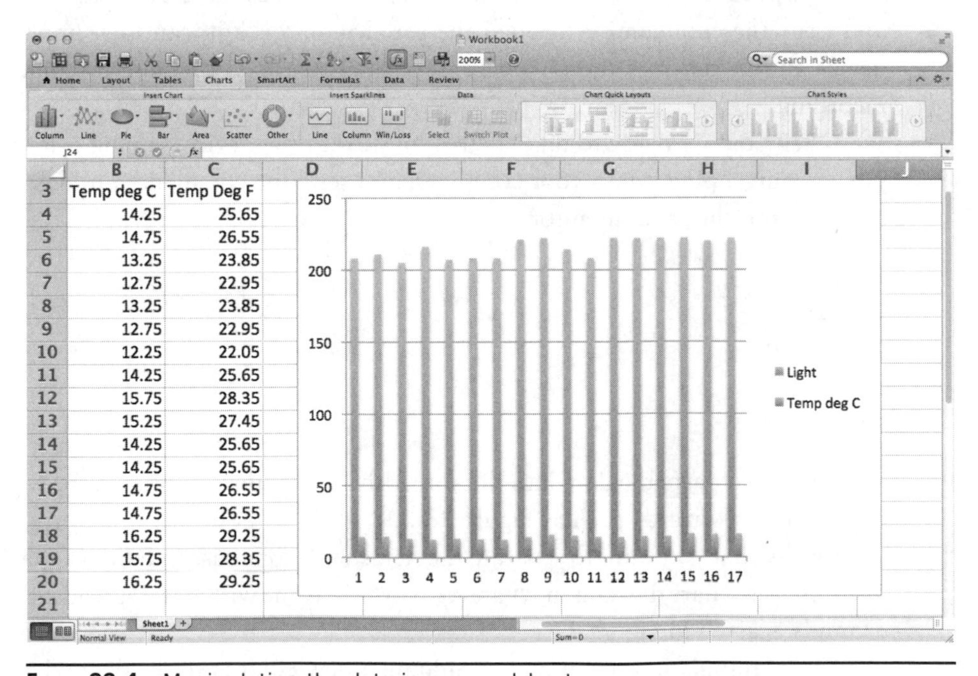

Figure 33-4 Manipulating the data in a spreadsheet.

just delete it after importing. Alternatively, and probably better, edit the sketch so that it only records and dumps the temperature data.

Summary

In Chapter 34, we will move away from sensing and make a timer to control a high-brightness 12-V LED lamp using a metal-oxide-semiconductor field-effect transistor (MOSFET).

CHAPTER 34
Timer-Controlled Lamp

Difficulty: ★★★	Cost guide: $30

Although displaying something on an LCD module is pretty straightforward, it can be surprisingly difficult to write the sketches for comparatively simple projects such as a timer that need to vary what is displayed. This is such a project. The hardware is easy, but the software a little more tricky. Figure 34-1 shows the project in action.

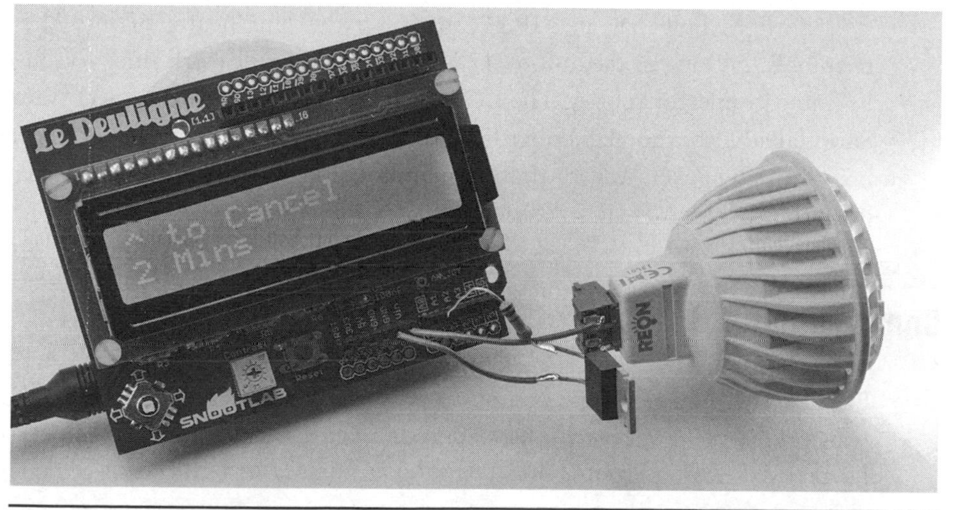

FIGURE 34-1 Timer-controlled lamp.

This is another project, like Chapters 29 and 30, that use the Deuligne LCD shield. It allows you to use the joystick switch to set a time for which the light should be lit. The light in question is a 12-V, 5-W LED light bulb.

WARNING *The metal-oxide-semiconductor field-effect transistor (MOSFET) used in this project will easily power the 12 V, 5 W bulb and also would be fine with a 12 V, 10 W blub. However, under no circumstances should you try to use it to control 110 or 220 V alternating current (ac). It will not work and would be extremely dangerous. This project is strictly for low-voltage direct current (dc).*

Parts

To build this project, you will need the following:

Part	Quantity	Description	Appendix
	1	Snootlabs Deuligne LCD shield	M21
R1	1	1 kΩ resistor	R2
T1	1	N-channel power MOSFET	S7
	1	Two-way 0.2-inch screw terminal	H7
	1	12 V, 5 W LED light bulb	
	1	12 V, 1 A power supply	M3

The light bulb has an MR16 fitting, so the pins neatly fit into the screw terminals. As long as the bulb is 12 V, another sort will work fine; you just have to find a connector to fit it. The LCD shield that I used in this project is made by Snootlabs (www.snootlab.com). It is particularly useful for this project because it has through-socket headers that allow access to the pins. Most LCD shields do not offer this.

Construction

Because the only electronic components for this project are the MOSFET and a resistor, the components can be soldered directly to each other and the terminal block, as shown in Figure 34-2.

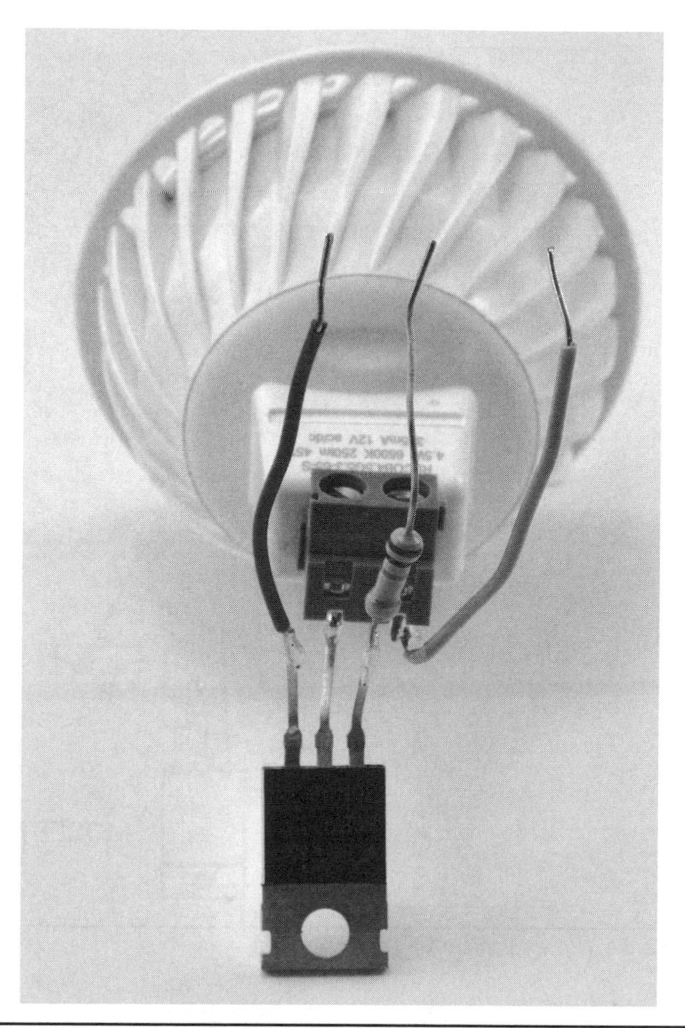

Figure 34-2 Soldering the components.

Referring to Figure 34-2, two short lengths of solid-core wire are soldered to the left-hand pin of the MOSFET and the right-hand pin on the terminal block. The middle pin of the MOSFET is connected to the left-hand pin of the terminal block and the resistor, with one leg shortened, which then connects to the right-hand lead of the MOSFET. Trim all three leads so that they are more or less the same length. They will plug into the sockets on the LCD shield. Figure 34-3 shows the wiring arrangement for the project.

The wire from the left-hand pin of the MOSFET goes to Arduino GND, the lead from the right-hand pin of the screw terminal goes to Arduino Vin, and the resistor lead goes to Arduino A1 on the LCD shield.

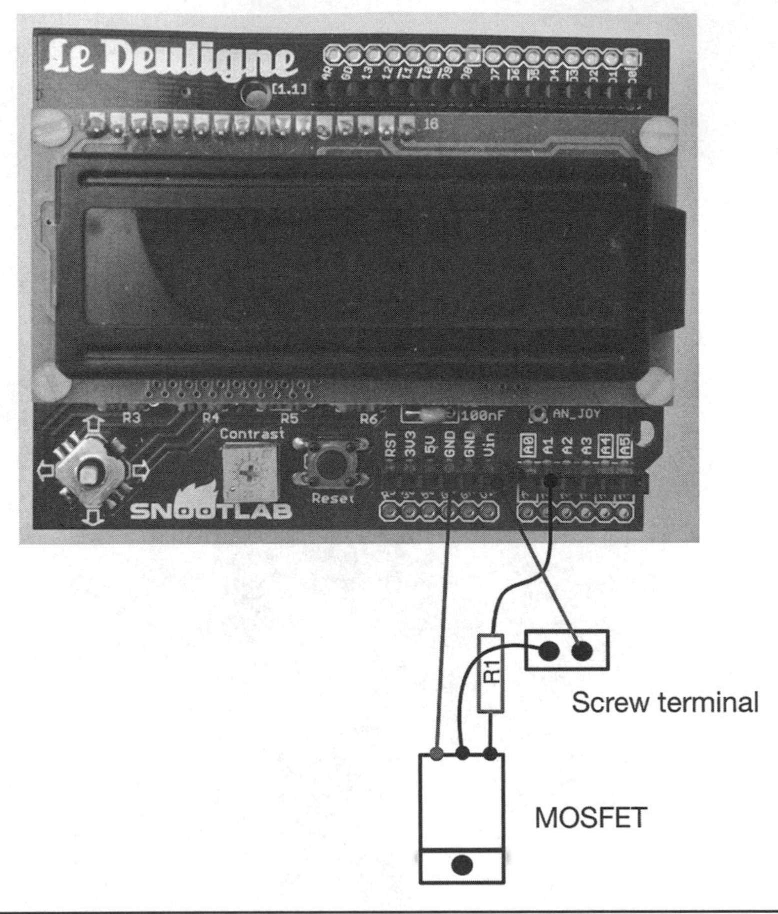

FIGURE 33-3 Wiring diagram for the timer light.

Software

The sketch for this project is ch_34_timer_lamp, and it requires a library for the LCD display shield. You can download the library from https://github.com/Snootlab/Deuligne/zipball/Arduino-1.0. This sketch can be in one of two modes. The user can either be setting the time delay or the timer can be timing. You can move between modes by interacting with the joystick keys on the display shield. The key to getting the software for a project like this working right is to model those modes or "states" in your code. A good way to do this is to use a `state` variable that can be set to different constant values for all its possible states. In this case, there are just two states, represented in the following constants:

```
const int STATE_SETTING = 0;
const int STATE_RUNNING = 1;
```

We will also define some constants representing the different values that the Deuligne library returns when the joystick switches are moved. These are

```
const int KEY_LEFT = 3;
const int KEY_RIGHT = 0;
const int KEY_UP = 1;
const int KEY_DOWN = 2;
```

We want to make it easy to change the delays available to chose from and the text that is displayed when selecting between these. To do this, we will define some constant arrays—`delays` to hold the delays in seconds and " to hold a corresponding set of labels to appear on the LCD screen.

```
const int numDelays = 5;
const long delays[] = {10, 30, 60, 120, 300};
static const char labels[numDelays][11] = {
  "10 Seconds",
  "30 Seconds",
  "1 Min     ",
  "2 Mins    ",
  "5 mins    " };
```

Yet more constants are used for the two messages that appear on the LCD screen telling the user what to do.

```
static const char settingPrompt[] = "< and >, v run";
static const char runningPrompt[] = "^ to Cancel ";
```

Next, we have some variables. There is the `state` variable itself and a `delayIndex` that holds the position of the currently selected delay option. The `long` variable `offTime` holds time in milliseconds after which the lamp should be turned off. We will see how this works later.

```
int state = STATE_SETTING;
int delayIndex = 0;
long offTime = 0;
Deuligne lcd;
```

The `setup` function performs the necessary initialization of the LCD shield and output pin. It also sets the initial message to appear on the LCD.

```
void setup()
{
  pinMode(outputPin, OUTPUT);
  lcd.init();
  lcd.clear();
  Serial.begin(9600);
  lcd.print(settingPrompt);
  changeDelay(0);
}
```

When you model states like this in a program, the mechanism for controlling the movement between states is described as a *state machine*. The way to implement a state machine is to have `if` statements in your `loop` function that test to see what state you are in and then performs the necessary steps, including checking for conditions that might result in changing to a different state. This makes the code quite a long.

```
void loop()
{
  int key = lcd.get_key();
  if (state == STATE_SETTING)
  {
    lcd.setCursor(0, 0);
    if (key == KEY_LEFT)
    {
      changeDelay(-1);
    }
    else if (key == KEY_RIGHT)
    {
      changeDelay(1);
    }
    else if (key == KEY_DOWN)
    {
      lcd.setCursor(0, 0);
      lcd.print(runningPrompt);
      offTime = millis() + delays[delayIndex] * 1000;
      digitalWrite(outputPin, HIGH);
      state = STATE_RUNNING;
    }
  }
```

```
  else if (state == STATE_RUNNING)
  {
    if (key == KEY_UP || millis() > offTime)
    {
      digitalWrite(outputPin, LOW);
      lcd.setCursor(0, 0);
      lcd.print(settingPrompt);
      state = STATE_SETTING;
    }
  }
}
```

As an example of this, the condition for handling the state `STATE_RUNNING` at the end of the function is the simpler example to follow. This checks to see if the "Up" button on the joystick has been pressed or, alternatively, that the correct amount of time has elapsed. If either of these is `TRUE`, then it changes state back to `STATE_SETTING` after changing the message that is displayed.

The utility function `changeDelay` both sets the `delayIndex` and updates the display.

```
void changeDelay(int direction)
{
  delayIndex += direction;
  if (delayIndex < 0) delayIndex = 0;
  if (delayIndex >= numDelays) delayIndex = numDelays;
  lcd.setCursor(0, 1);
  lcd.print(labels[delayIndex]);
  delay(500);
}
```

Using the Project

Because the project uses a 12 V lamp, a 12 V dc power supply is used to supply power to both the Arduino and the lamp. To use the project, select the delay that you want by using left and right movements of the joystick. When ready, nudge the joystick down to light the lamp for the period specified. You can cancel the timer, turning the lamp back off, by moving the joystick up.

Summary

In the next project, we will create an auto-ranging capacitance meter. Something that will come in very handy if you have some capacitors to test.

CHAPTER 35

Autoranging Capacitance Meter

Difficulty: ★★	Cost guide: $35

This project uses an LCD shield to display the value of a capacitor attached to test sockets (Figure 35-1). The meter will measure the value of capacitors greater than 1 nF up to several hundred microfarads.

Figure 35-1 Autoranging capacitance meter.

Parts

To build this project, you will need the following:

Part	Quantity	Description	Appendix
	1	Snootlabs Deuligne LCD shield	M21
R1	1	1 kΩ, ¼ W resistor	R2
R2	1	1 MΩ, ¼ W resistor	R7
	1	Right-angle header pins × 9	H12
	1	Header sockets, 2 × 4 way	H2
	1	Stripboard, 10 strips × 14 holes	H5
		Capacitors to test: 1 nF to 100 μF	

The LCD shield that I used for this project is made by Snootlabs (www.snootlab.com). It is particularly useful for this project because it has through-socket headers that allow access to the pins. Most LCD shields do not offer this.

Construction

This project uses just two resistors to measure the capacitance, and if you would rather use, say, crocodile clips than the header sockets, you could solder the components directly to normal header pins. Figure 35-2 shows the layout of the stripboard. There are no cuts to make in the strip, and the only slightly tricky part is attaching the right-angle header pins.

Step 1: Cut the Stripboard to Size

Stripboard is a kind of ready-made printed circuit board (PCB) for prototyping. You are unlikely to find a piece that is just the right size, so the first step is to cut your board so that there are eight strips each of 55 holes (Figure 35-3). It does not matter if it is a bit larger than this.

Stripboard is not easy to cut. It is better to score it on both sides and then break it over the edge of your worktable. Do not try to break it between holes. Break it along a row of holes.

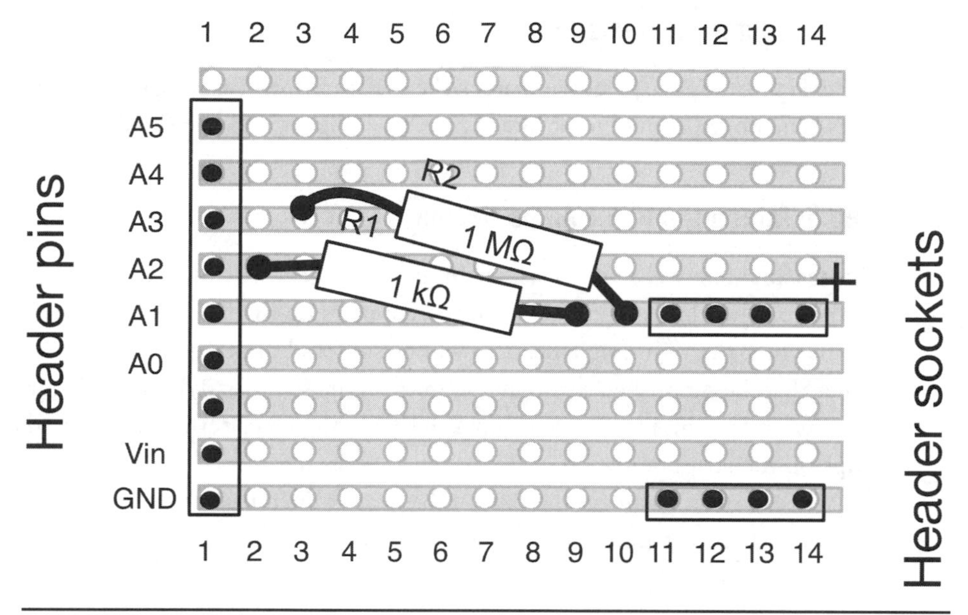

FIGURE 35-2 Stripboard layout.

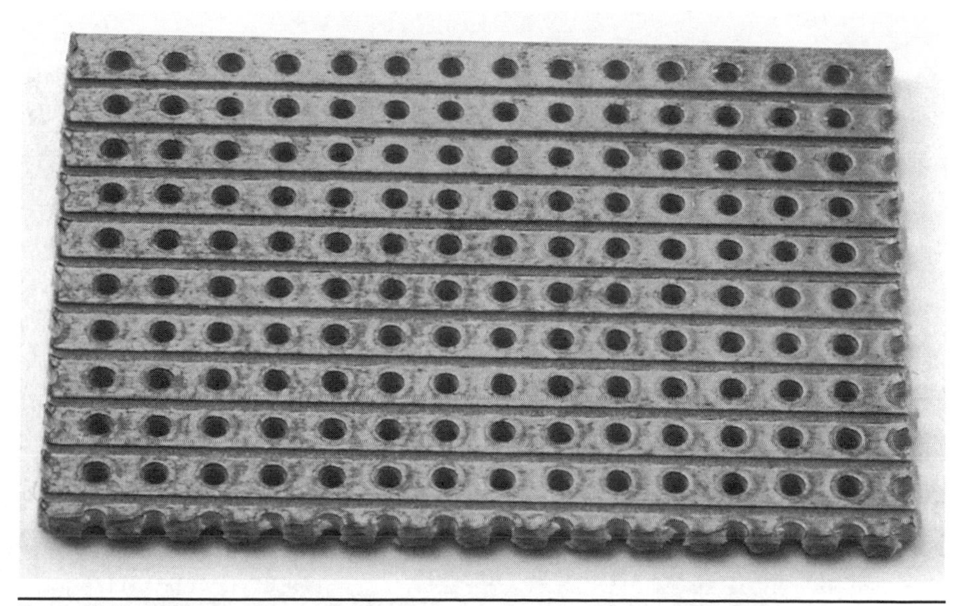

FIGURE 35-3 Stripboard cut to size.

Step 2: Solder the Resistors

Using Figure 35-2 as a guide, solder the resistors into place. Cut off the excess leads. The board should look like Figure 35-4 when you have finished.

Step 3: Attach the Right-Angle Header

The reason that we have to use right-angle headers in this project rather than normal pin headers is that we want the copper side of the stripboard to be face down. Thus, the right-angle headers will be soldered directly to the copper tracks rather than being pushed through the holes on the stripboad. Refer to Figure 35-1 to see what I mean. Figure 35-5 shows the right-angle header being soldered to the strips on the stripboard.

Make sure that you get the pins of the header lined up correctly relative to the resistors. Refer to Figure 35-2 to get this right.

Step 4: Attach the Header Sockets

You can now attach the header sockets as shown in Figure 35-6.

The header sockets will need to be cut from a longer strip of header sockets.

Figure 35-4 Stripboard with resistors attached.

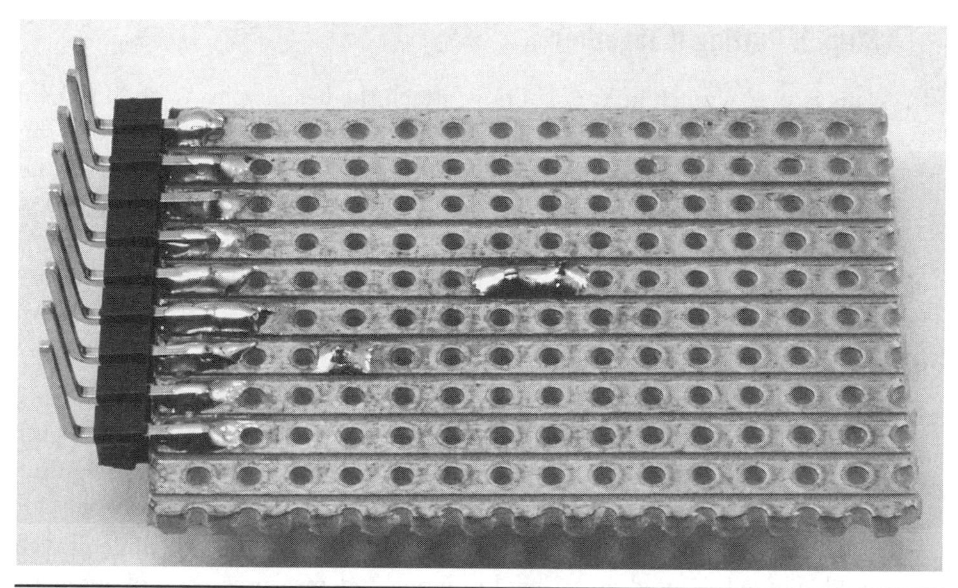

Figure 35-5 Soldering the right-angle header to the stripboard.

Figure 35-6 Soldering the header sockets.

Step 5: Putting It Together

This is pretty much it. You can now attach the header pins to the LCD shield as shown in Figure 35-1. It fits over the whole of the analog row of headers with two sockets from the power header next to it. There is also one unused pin between the two sets of header sockets on the LCD shield. You now just need to load up the software and find a capacitor or two to test.

Software

The sketch for this project is ch_35_capacitance_meter, and it requires a library for the LCD shield. You can download the library from https://github.com/Snootlab/Deuligne/zipball/Arduino-1.0.

After importing the libraries, constants are defined for the three pins that are used. Pin A1 is used as an analog input and measures the voltage across the capacitor as it is charged up through either the 1 kΩ resistor R1 or the 1 MΩ resistor R2. These two resistors are controlled by pins A2 and A3. If the pin for a particular resistor is set to be an input, then the resistor takes no part in the charging or discharging of the capacitor. However, if it is set to be an output and low, the capacitor will discharge through it. If it is an output and high, then the capacitor will charge through it.

```
const int stepHighCPin = A2;
const int stepLowCPin = A3;
const int sensePin = A1;
```

Constants are also used for the values of the two resistors.

```
const float R1 = 1000.0;
const float R2 = 1000000.0;
```

The `threshold` constant is the analog reading value (0 to `1,023`) at which the capacitor is deemed to be charged sufficiently to record the amount of time it took to charge using R1 or `R2`. This is then the time constant and is 63 percent of the full 5-V reading.

```
const int threshold = 645; // 63%
```

This project relies on timing how long it takes to charge up the test capacitor to a certain level. This requires the analog measurement to be as fast as possible. The following lines of code speed up the measurement from around 9,000 readings per second to 58,000 readings per second:

```
// speed up ADC
const byte PS_128 = (1 << ADPS2) | (1 << ADPS1) | (1 << ADPS0);
const byte PS_16 = (1 << ADPS2);
```

The LCD display object is defined using

```
Deuligne lcd;
```

The `setup` function sets both resistor control pins to be inputs to start with and then initializes the display with the message `Capacitance` on the top line of the display.

```
void setup()
{
  pinMode(stepHighCPin, INPUT);
  pinMode(stepLowCPin, INPUT);
  lcd.init();
  lcd.clear();
  lcd.print("Capacitance");
}
```

The `loop` function first takes a reading for high-value capacitors using the `readC` function (we will come to this in a moment). If this reading is too low, it effectively switches ranges, becoming 1,000 times more sensitive, and takes another reading. It then calls `display` to display the reading on the LCD and then waits for half a seconds before starting again.

```
void loop()
{
  float highRange = readC(R1, stepHighCPin, stepLowCPin);
  float reading = highRange;
  if (highRange < 1.0F)
  {
    float lowRange = readC(R2, stepLowCPin, stepHighCPin);
    reading = lowRange;
  }
  display(reading);
  delay(500);
}
```

The `readC` function first calls discharge to empty the capacitor of charge and then times how long it takes to charge the capacitor through the pin specified (stepPin). The value of capacitance in microfarads can be calculated as the time taken in microseconds divided by the value of the resistor used in ohms.

```
float readC(float r, int stepPin, int unusedPin)
{
  discharge();
  long t0 = micros();
  charge(stepPin, unusedPin);
  long t1 = micros();
  float T = float(t1 - t0);
  float C = T / r;
  return C;
}
```

We want discharging to be as quick as possible, so discharging is always done with R1, the 1 kΩ resistor. The discharge function, having enabled R1, waits until the analog reading from the capacitor reaches 0.

```
void discharge()
{
  pinMode(stepLowCPin, INPUT);
  pinMode(stepHighCPin, OUTPUT);
  digitalWrite(stepHighCPin, LOW);
  while (analogRead(sensePin) > 0) {};
}
```

When charging the capacitor being tested, the appropriate resistor pin is set to be an output and HIGH. The unused resistor control pin is set to be an input. The charge function simply waits in a loop until it reaches the threshold.

```
void charge(int stepPin, int unusedPin)
{
  pinMode(unusedPin, INPUT);
  pinMode(stepPin, OUTPUT);
  digitalWrite(stepPin, HIGH);
  while (analogRead(sensePin) < threshold) {};
}
```

There are three possible paths for what will be displayed on the LCD, controlled by the display function. If the reading is below 0.001 (1 nF), then the display reads Too Low. If it is greater than 0.8 μF, then it displays the reading in microfarads; otherwise, it displays it in nanofarads.

```
void display(float reading)
{
  if (reading < 0.001)
  {
    clearRow();
    lcd.print("Too Low");
  }
  else if (reading > 0.8)
  {
    clearRow();
    lcd.print(reading);
    lcd.print(" uF");
  }
  else
  {
    clearRow();
    lcd.print(reading * 1000.0);
    lcd.print(" nF");
  }
}
```

The function clearRow is used to blank out the bottom line of the display before displaying the new value.

```
void clearRow()
{
  lcd.setCursor(0, 1);
  lcd.print("                ");
  lcd.setCursor(0, 1);
}
```

Using the Project

This is not a very accurate capacitance meter, but it will help in identifying capacitors or in checking out suspect devices.

Summary

This technique for measuring capacitance could be easily adapted to measure resistance if fixed-value capacitors are used and the resistor becomes the variable element. In the next and final chapter, we will make an Arduino-based Geiger counter.

CHAPTER 36

Geiger Counter

Difficulty: ★★★★	Cost guide: $30

This project uses a low-cost Geiger-Müller tube to detect gamma and beta radiation (Figure 36-1). If you do not have a radiation source, then the project is still interesting in that it will also detect cosmic rays.

WARNING *This project generates the 450 V that the Geiger-Müller tube requires to operate. The voltage is generated from the Arduino's USB power supply and is extremely low current (and therefore relatively low risk). However, you*

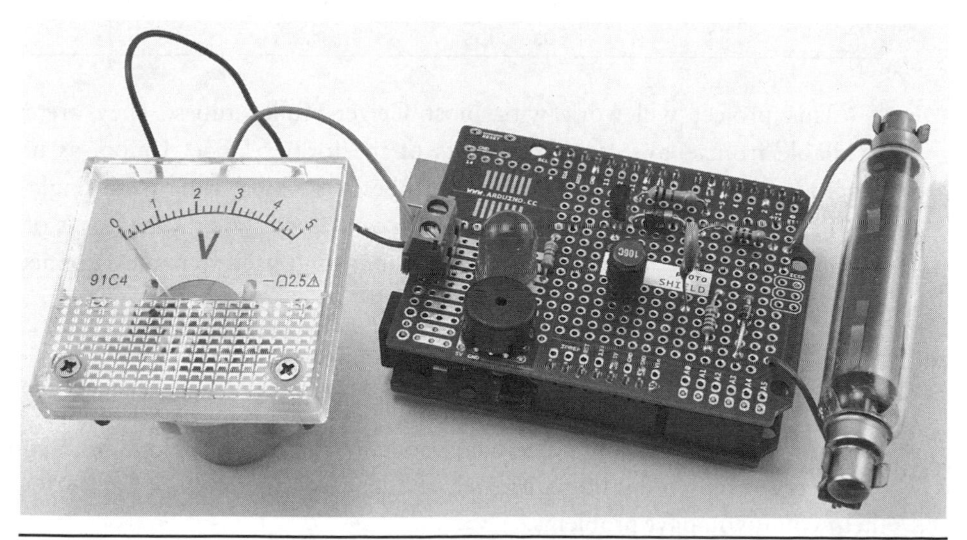

FIGURE 36-1 Geiger counter.

should take special care with this project because much of the Protoshield will be at 400 V and can give you a surprising jolt.

Parts

To build this project, you will need the following:

Part	Quantity	Description	Appendix
	1	Protoshield and header pins	A3, H1
	1	Geiger-Müller tube	M23
R1	1	1 kΩ, ¼ W resistor	R2
R2	1	1 MΩ, ¼ W resistor	R7
R3	1	270 Ω, ¼ W resistor	R1
R4	1	47 kΩ, ¼ W resistor	R8
C1	1	100 nF, 1,000 V	C7
L1	1	10 mH inductor	C8
D1	1	1N4937 diode	S14
D2	1	5 V Zener diode	S15
D3	1	Red LED	S8
T1	1	MPSA44 transistor	S16
	1	Piezo sounder	C3
	1	5 V analog meter	M19
	1	Two-way 0.2-inch screw terminal	H7
	1	Fuse clips	

This project will work with most Geiger-Müller tubes. They are readily available from eBay, often from parts of the former Soviet Union, as new-old stock. Finding a clip connector for the tubes is trickier. Mine were single-ended PCB fuse holders, although in a pinch you can simply wrap a few turns of solid-core wire around each end and then wrap it around itself to make a connection.

Construction

This project requires quite a few components and is one of the more complex projects to build in this book. As such, the schematic diagram of Figure 36-2 may help you if you have problems.

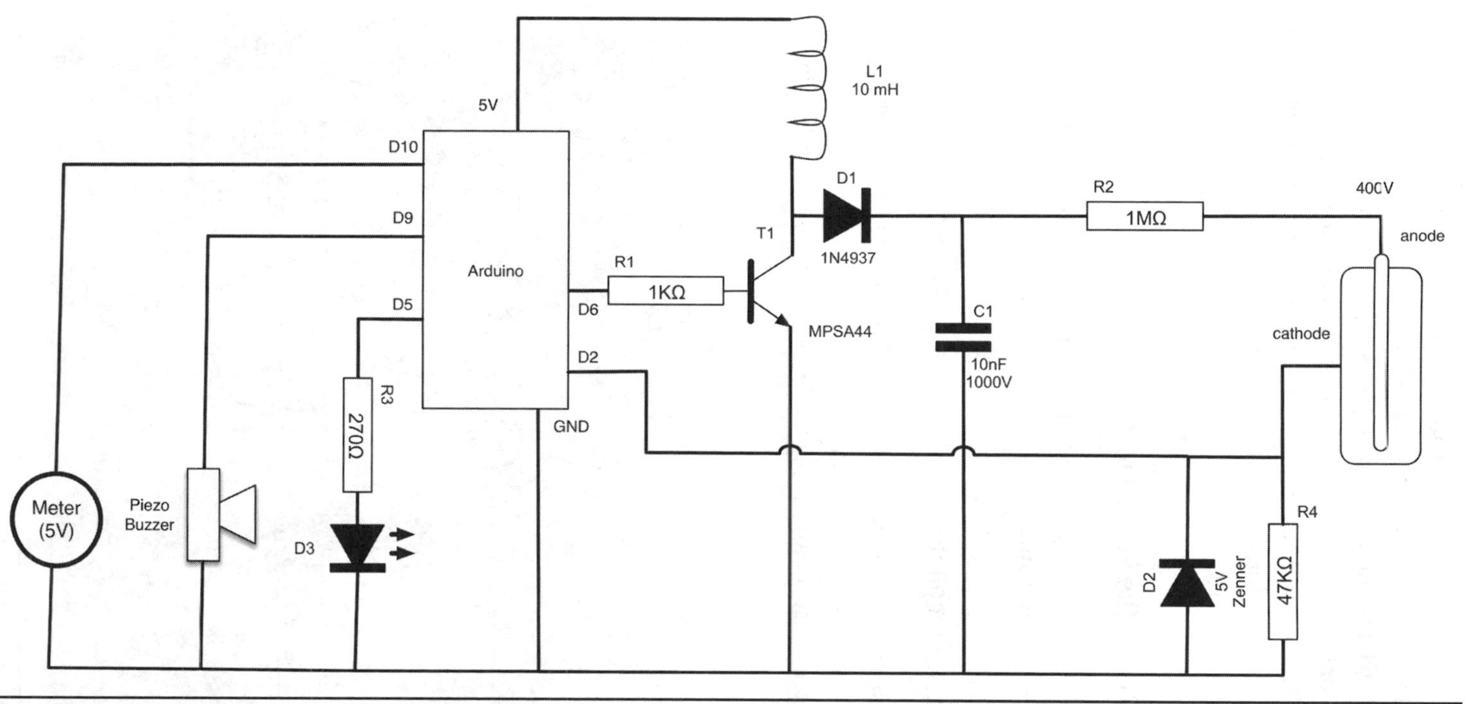

FIGURE 36-2 Schematic diagram for the Geiger counter.

T1, D1, R1, and L1 are driven by a squarewave signal at 1.5 kHz to generate the 450 V required by the anode of the tube. These are therefore high-voltage components, as is C1, which smoothes the 450-V supply. When radiation passes through the tube, it ionizes the gas inside, creating a pulse at the cathode. The magnitude of this pulse will vary, but some of the time will be higher than 5 V, so the Zener diode is used to protect the Arduino input. Figure 36-3 shows the Protoshield layout for the board.

Step 1: Attach the Pins to the Protoshield

This project does not use the analog section of the Arduino, so this section on pin headers can be omitted if you want.

Step 2: Solder the Low Components

Next, solder the resistors and diodes onto the Protoshield. Leave the leads untrimmed so that they can be used to join up the connections on the underside of the board. When the resistors are in place, the board should look like Figure 36-4.

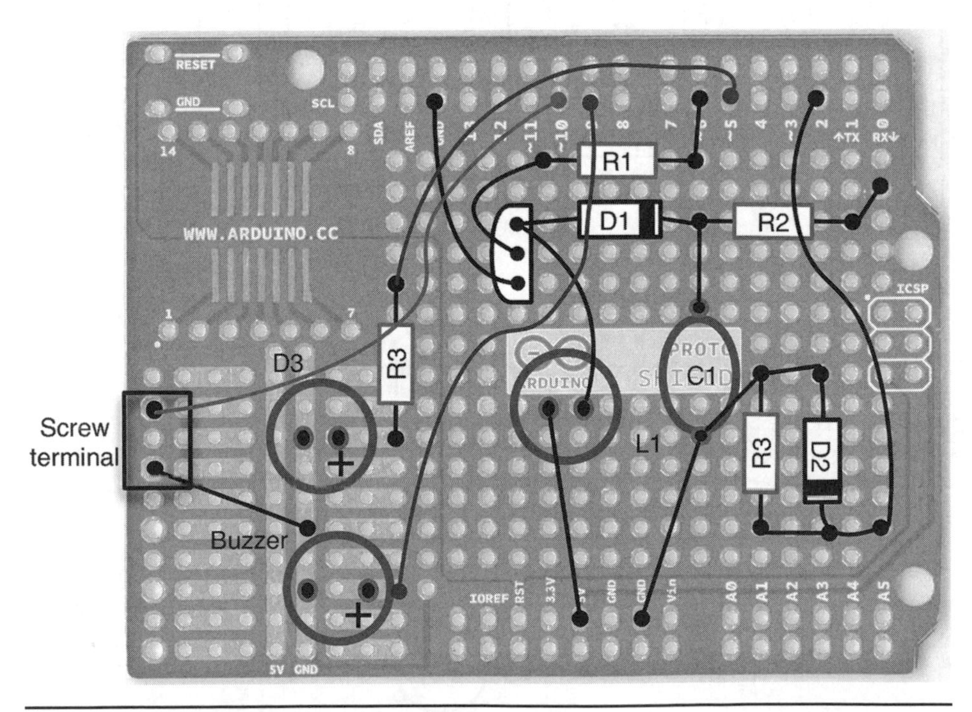

FIGURE 36-3 Protoshield layout for the Geiger counter.

FIGURE 36-4 Protoshield with resistors and diodes attached.

Note that the diodes both have a bar at one end. Make sure that these are the right way around, as shown in Figure 36-3.

Step 3: Solder the Remaining Components

The remaining components can now be soldered to the board, starting with the lowest components. The longer lead of the LED is the positive lead, and the buzzer also may have a polarity marked on it. The inductor can go either way around, but make sure that the transistor has the curved edge as shown in Figure 36-3. You can see the appearance of the top of the board in Figure 36-1.

Step 4: Solder the Underside of the Board

Figure 36-5 shows the connections from the point of view of the underside of the board.

In most cases, the connections can be made by bending over the component leads and soldering them together. You will need some lengths of insulated solid-core wire to make some of the connections. When all the links have been made, the board should look like Figure 36-6.

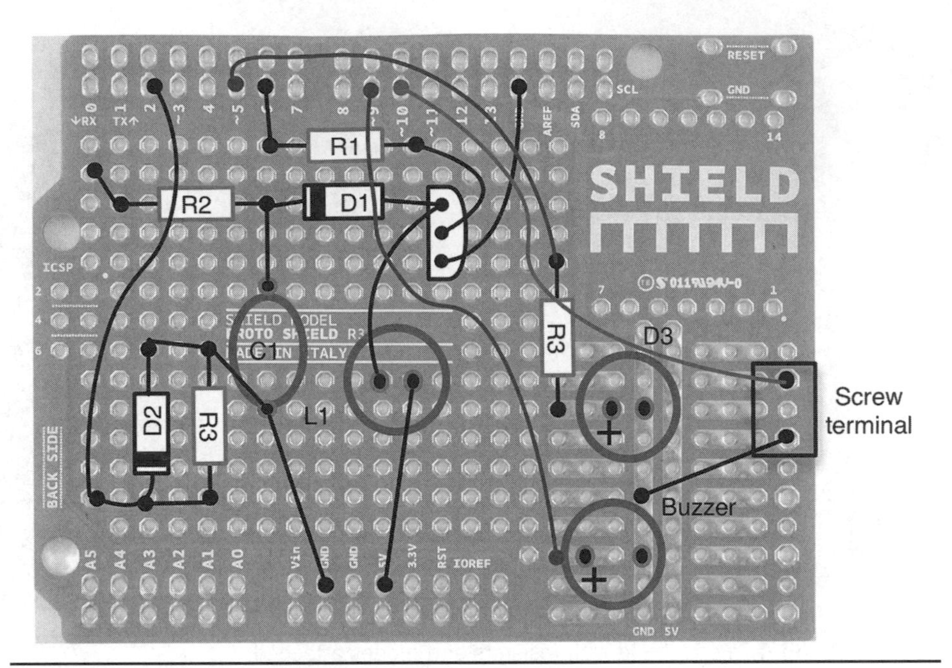

FIGURE 36-5 Wiring from the underside of the board.

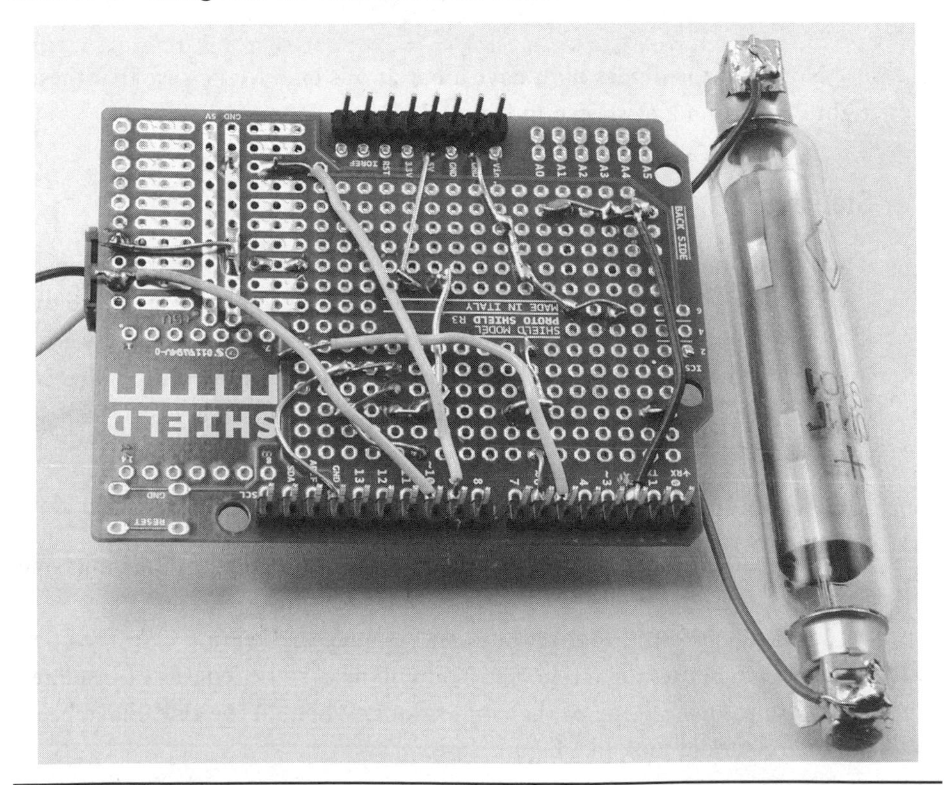

FIGURE 36-6 Underside of the board.

Step 5: Attach the Tube and Meter

Use two lengths of solid-core wire connected to fuse clips or wrapped around the tube ends and then carefully soldered. Note that the tube will be marked to indicate the anode, usually with a red dot on the glass. You can also attach the voltmeter to the two screw terminals.

Software

The sketch for this project can be found in the file ch_36_geiger_counter. It does not require the installation of any libraries.

The sketch starts with the definition of some constants for pin definitions.

```
const int powerPin = 6;
const int ledPin = 5;
const int buzzerPin = 9;
const int meterPin = 10;
const int sensePin = 2;
```

Next, we have a constant `smoothingFactor` that controls the smoothing of the value for the number of counts per minute that is displayed on the voltmeter. Three variables are also defined.

```
long cancelTime = 0;
long lastEventTime = 0;
long cpm = 0;
boolean canceled = true;
```

The variable `cancelTime` is used after an event has been detected by the tube to schedule a time at which the beep of the buzzer is to stop and the LED is to turn back off. The Boolean variable `canceled` is a flag indicating whether the buzz or LED being on has canceled to speed up the processing in the main loop. The variable `lastEventTime` is used to calculate the time between events and hence the number of counts per minute (`cpm`).

The `setup` function does little more than set up the pin modes and start a 1.5-kHz tone playing on the `powerPin` to generate the high-voltage supply.

```
void setup()
{
  pinMode(ledPin, OUTPUT);
  pinMode(buzzerPin, OUTPUT);
```

```
  pinMode(sensePin, INPUT);
  tone(powerPin, 1500);
  Serial.begin(9600);
}
```

The main loop is as follows:

```
void loop()
{
  if (digitalRead(sensePin))
  {
    event();
  }
  else if (! cancelled && millis() > cancelTime)
  {
    digitalWrite(ledPin, LOW);
    analogWrite(buzzerPin, 0);
    cancelTime = millis();
    cancelled = true;
  }
}
```

The main loop repeatedly reads the sensePin looking for a pulse that indicates an event. This loop runs very quickly until an event is detected, so even the shortest pulses will be detected. An alternative implementation would be to use interrupts, but this suffers from the disadvantage of interfering with the operation of the tone command.

If an event is detected, the event function is called. Otherwise, if the number of milliseconds elapsed is greater than the time set for the cancellation of buzzing, then buzzing is canceled and the LED turned off. The event function is as follows:

```
void event()
{
  long eventTime = millis();
  digitalWrite(ledPin, HIGH);
  analogWrite(buzzerPin, 127);
  cancelled = false;
  cancelTime = eventTime + 10;

  long elapsed = eventTime - lastEventTime;
  long newCpm = 60000 / elapsed;
```

```
  cpm = (newCpm * smoothingFactor / 100)
          + (cpm * (100 - smoothingFactor) / 100);
  // 500 cpm fsd
  int meterValue = cpm * 255 / 500;
  if (meterValue > 255) meterValue = 255;
  analogWrite(meterPin, meterValue);
  Serial.println(meterValue);
  lastEventTime = eventTime;
  delay(100);
}
```

The `event` function logs the time of the event in the variable `eventTime` and then turns on the LED and buzzer, at the same time scheduling the cancellation time as 10 milliseconds in the future. It then calculates the counts per minute using the elapsed time since the last event. It also calculates an analog output value for the meter based on a full-scale deflection as being 500 counts per minute.

The 100-millisecond delay at the function prevents a long pulse counting as more than one event. This does mean, however, that two events very close together will result in only one event being registered.

Using the Project

Upload the sketch before attaching the Protoshield. If everything is working okay, you should get a beep and a buzz every few seconds as a background count. The meter is only really of much use if you have a radiation source that you can use to increase the number of events being read by the Geiger counter.

Summary

This is the final project of this book, and I hope that it and the other projects here will inspire you to create some Arduino projects of your own. If you want to learn more about programming Arduino, then you might like to consider some of my other books on Arduino, including *Programming the Arduino: Getting Started with Sketches* and the more advanced *Programming Arduino Next Steps*.

APPENDIX

Component Buying Guide

All the components used in this book are easily available from suppliers on the Internet. Alternatively, entering the description of a component from the following tables will usually yield results on eBay.

Useful suppliers that ship worldwide include

- Sparkfun: www.sparkfun.com
- Adafruit: www.adafruit.com
- Mouser: www.mouser.com
- Digikey: www.digikey.com
- Newark: www.newark.com

In the United Kingdom there are

- Maplins: www.maplins.com
- CPC: http://cpc.farnell.com

If you are looking for kits, then MonkMakes (www.monkmakes.com) has kits designed specifically for this book.

Arduino

Arduino products are very widely available; all the major suppliers, including all those mentioned earlier, sell Arduino. But shop around because prices do vary a little.

Book Code	Description	Source
A1	Arduino Uno R3	
A2	Arduino Leonardo	
A3	Protoshield bare printer circuit board (PCB)	eBay

Hardware

Book Code	Description	Source
H1	Strip of 0.1-inch header pins	Sparkfun: PRT-00116 Adafruit: 392
H2	Strip of 0.1-inch header sockets	Sparkfun: PRT-00115 Adafruit: 598
H4	3.5-mm PCB header socket	Sparkfun: PRT-08032[a] Newark: 24M4876 CPC: AV15106
H5	Stripbopard	Newark: 96K6336
H6	Jumper wires, male to female	Sparkfun: PRT-09385 Adafruit: 825
H7	Two-way screw terminal, 0.2 inch	Sparkfun: PRT-08432
H8	Stackable shield header set	Sparkfun: PRT-11417
H9	Jumper wires, female to female	Adafruit: 266 Sparkfun: PRT-08430
H10	8-pin DIL integrated-circuit (IC) holder	Sparkfun: PRT-07937
H11	Jumper wires, male to male	Adafruit: 758 Sparkfun: PRT-08431
H12	Right-angle header pins × 9	Adafruit: 1540 Sparkfun: PRT-00553
H13	Three-way screw terminal, 0.2 inch	Sparkfun: PRT-08433

[a]Sparkfun design has a different pin-out from the socket used, which is the Newark/CPC model. The type of socket used is also usually available on eBay.

Modules and Sheilds

Book Code	Description	Source
M1	Sparkfun Tilt-a-Whirl tilt sensor	Sparkfun: SEN-10621
M2	12 V light-emitting diode (LED) module	eBay
M3	12 V, 2 A power adapter	Sparkfun: TOL-11296 Adafruit: 798
M4	Adafruit bicolor LED matrix	Adafruit: 902
M5	Adafruit real-time clock (RTC) module	Adafruit: 264
M6	Adafruit four-digit, seven-segment 5.6-inch LED module	Adafruit: 878a
M7	Color recognizer module	eBay
M8	Adafruit PN532 NFC/RFID shield	Adafruit: 789
M9	12 V door latch	Farnell: 3541071
M10	Adafruit four-digit, seven-segment 1.2-inch LED module	Adafruit: 1270
M11	HC-SR-04 ultrasonic range finder	eBay
M12	5 V Arduino laser module	eBay
M13	Arduino screw shield	Freetronics: SH-TERMINAL (106)
M14	Adafruit ADXL335 accelerometer module	Adafruit: 163
M15	TEA5767 breakout PCB	MonkMakes: 00020
M16	Arduino Ethernet shield	Adafruit: 201 Sparkfun: DEV-09026
M17	PowerSwitch tail	Adafruit: 268 Sparkfun: COM-10747
M18	DHT11 temperature and humidity sensor	Adafruit: 386
M19	5 V panel meter	Sparkfun: TOL-10285
M20	MQ4 methane gas sensor	Sparkfun: SEN-09404
M21	Snootlabs Deuligne liquid-crystal display (LCD) shield	http://snootlab.com/lang-en/snootlab-shields/135-deuligne-en.html
M22	Adafruit ultimate GPS breakout v3	Adafruit: 746
M23	450 V Geiger-Müller tube	eBay
M24	Adafruit 1.8-inch TFT shield	Adafruit: 802
M25	Sparkfun MP3 player shield	Sparkfun: DEV-10628

[a]*Also available in other colors.*

Resistors

An alternative to buying individual resistors is to buy a kit such as the COM-10969 from Sparkfun or a similar kit offered by Monkmakes.com. The power ratings are the minimum power, so 1/2-W resistors are also fine if you cannot find ¼-W resistors.

Book Code	Description	Source
R1	270 Ω, ¼ W resistor	Mouser: 273-270-RC
R2	1 kΩ, ¼ W resistor	Mouser: 293-1K-RC
R3	100 Ω, ¼ W resistor	Mouser:293-100-RC
R4	4.7 MΩ, ¼ W resistor	Mouser: 293-4.7M-RC
R5	200 kΩ, ¼ W resistor	Mouser: 293-200K-RC
R6	Photoresistor (1 kΩ)	Adafruit: 161 Sparkfun: SEN-09088
R7	1 MΩ, ¼ W resistor	Mouser: 293-1M-RC
R8	47 kΩ, ¼ W resistor	Mouser: 293-47K-RC
R9	10 kΩ trimpot	Adafruit: 356 Sparkfun: COM-09806
R10	10 MΩ, ¼ W resistor	Mouser: 293-10M-RC
R11	20 kΩ, ¼ W resistor	Mouser: 293-20K-RC

Resistors have stripes on them that tell you their value. Each color has a value, as shown in the following table:

Black	0
Brown	1
Red	2
Orange	3
Yellow	4
Green	5
Blue	6
Violet	7
Gray	8
White	9
Gold	1/10
Silver	1/100

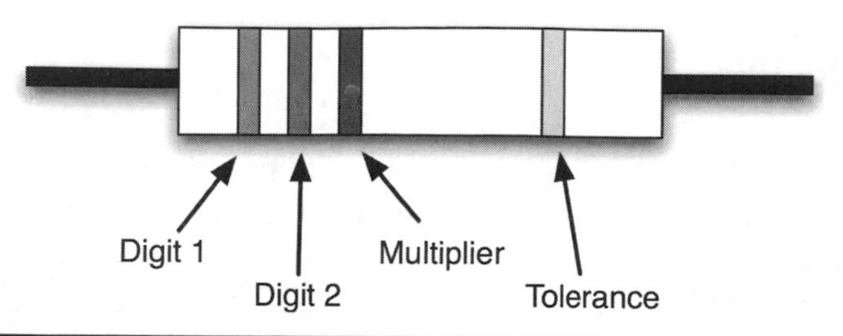

FIGURE A-1 Resistor stripes.

Gold and silver, as well as representing the fractions 1/10 and 1/100, are also used to indicate how accurate the resistor is, so gold is ±5 percent and silver is ±10 percent. There will generally be three bands together starting at one end of the resistor, a gap, and then a single band at the other end of the resistor. The single band indicates the accuracy of the resistor value. Because none of the projects in this book requires very accurate resistors, there is no need to select your resistors on the basis of accuracy.

Figure A-1 shows the arrangement of the colored bands. The resistor value uses just the three bands. The first band is the first digit; the second, the second digit; and the third, multiplier band is how many zeros to put after the first two digits. Thus, a 270-Ω (pronounced "Ohm") resistor will have a first digit 2 (red), a second digit 7 (violet), and a multiplier of 1 (brown). Similarly, a 10-kΩ resistor will have bands of brown, black, and orange (1, 0, and 000).

Semiconductors

Light-emitting diodes (LEDs) are cheap and plentiful on eBay, especially if you do not mind waiting for them to arrive from China.

Book Code	Description	Source
S1	5-mm red LEDs	Digikey : 751-1118-ND Mouser: 941-C503BRANCY0B0AA1
S2	5-mm green LEDs	Digikey: 365-1186-ND Mouser: 941-C503TGANCA0E0792
S3	5-mm blue LEDs	Digikey: C503B-BCS-CV0Z0461-ND Mouser:78-TLHB4201
S4	5-mm yellow LEDs	Digikey:365-1190-ND Mouser: 941-C5SMFAJSCT0U0342
S5	5-mm white LEDs	Digikey : C513A-WSN-CV0Y0151-ND Mouser: 941-C503CWASCBADB152
S6	2N7000 MOSFET	Digikey: 2N7000TACT-ND Mouser: 512-2N7000
S7	Power MOSFET	Sparkfun: COM-10213 Digikey : FQP30N06L-ND Mouser: 512-FQP30N06
S8	10-mm red LED	Adafruit: 845 Sparkfun:COM-10632
S9	RGB common-cathode LED	Sparkfun: COM-09264
S10	2N3904 NPN bipolar transistor	Sparkfun: COM-00521
S11	1N4001 diode	Adafruit: 755 Sparkfun: COM-08589
S12	MSGEQ7 IC	Sparkfun: COM-10468
S13	TMP36 temperature sensor	Adafruit: 165 Sparkfun: SEN-10988
S14	1N4937 diode	Mouser: 512-1N4937
S15	5-V Zener diode	Mouser: 610-1N751A
S16	MPSA44 transistor	Mouser: 863-MPSA44 Farnell: 1574391

Other Components

Book Code	Description	Source
C1	Large push switch	Sparkfun: COM-09178
C2	Click switch	Sparkfun:COM-09190 Adafruit: 367
C2	Kaypad	Sparkfun: COM-08653
C3	Piezo buzzer	Sparkfun: COM-07950 Adafruit: 160
C4	5-V ("sugar cube") relay	Mouser: 893-833H-1C-S-5VDC
C5	100-nF capacitor (low voltage)	Adafruit: 753 Sparkfun: COM-08375 Mouser: 75-1C10X7R104K050B
C6	33-pF capacitor (check value)	Mouser: 581-AR155A330K4R
C7	100-nF capacitor (1,000 V)	Mouser: 5989-1KV0.1-F
C8	10-mH inductor	Mouser: 652-RLB1014-103KL

Index

References to figures are in italics.